T0156097

SpringerBriefs in Mathematics

SpringerBriefs in Mathematics showcases expositions in all areas of mathematics and applied mathematics. Manuscripts presenting new results or a single new result in a classical field, new field, or an emerging topic, applications, or bridges between new results and already published works, are encouraged. The series is intended for mathematicians and applied mathematicians.

More information about this series at http://www.springer.com/series/10030

Anatoliy Swishchuk

Change of Time Methods
in Quantitative Finance

 Springer

Anatoliy Swishchuk
Department of Mathematics
and Statistics
University of Calgary
Calgary, AB, Canada

ISSN 2191-8198   ISSN 2191-8201 (electronic)
SpringerBriefs in Mathematics
ISBN 978-3-319-32406-7 ISBN 978-3-319-32408-1 (eBook)
DOI 10.1007/978-3-319-32408-1

Library of Congress Control Number: 2016936745

Mathematics Subject Classification (2010): 60H10; 60G44; 60J75; 91B28; 91B74

Printed on acid-free paper

This Springer imprint is published by Springer Nature
The registered company is Springer International Publishing AG Switzerland

To my late parents

Preface

"Time changes everything except something within us, which is always surprised, by change"—Thomas Hardy.

The following book is devoted to the history of change of time methods (CTMs), the connection of CTMs to stochastic volatilities and finance, and many applications of CTMs. One may consider this book as a brief introduction to the theory of CTMs and as a handbook that can be used to apply to multiple real-life problems. As Winston Churchill once said, "...I only read for pleasure or for profit"; similarly, some may read this book for pleasure, some for profit (see disclaimer below!), and some for both. My intention is to satisfy all the readers who enjoy change of time methods, with reference to stochastic volatilities and quantitative finance. There is one book that resembles mine partly, this being Change of Time and Change of Measure by O. Barndorff-Nielsen and A. Shiryaev (World Scientific, 2010). The difference between the following book and the latter book is that this book focuses more on applications and presents some novel models (e.g. the delayed version of the Heston model) that are not covered by the monograph of Barndorff-Nielsen and Shiryaev. To some extent, someone may consider this book as a useful complement to the latter monograph. I hope that this book will attract a wide audience: from graduate students and quants to researchers in mathematical and quantitative finance and also to practitioners in finance and energy areas.

Calgary, AB, Canada   Anatoliy Swishchuk
February 2016

Disclaimer

We make no guarantees that the approaches contained in this book are free of error or that they will meet your requirements for any particular application. If you do use the approaches present in the following book, please do so at your own risk. We disclaim all liability for direct or consequential damages resulting from your use of these approaches.

Acknowledgements

I would greatly like to thank many of my colleagues and graduate students with whom I discussed or obtained some results presented in this book. I would also like to thank all the participants of the "Lunch at the Lab" finance seminar at our Department of Mathematics and Statistics at the University of Calgary, where all the results were first presented and tested.

Many thanks go to my family, my wife Mariya, my son Victor, and my daughter Julianna (who found and recommended the picture in the Preface), whose continuous support encouraged me on writing and creating.

I would like to especially thank my daughter Julianna for editing the grammar and vocabulary of the book. I greatly appreciate it.

I wish to thank Donna Chernyk (editor) and Suresh Kumar (TeX support team) from Springer USA for their continuous support and help during the book's preparation.

My final thanks go to many referees who reviewed not only sample chapters initially but also the whole final book. Their remarks, suggestions, and comments have definitely improved the present book. The remaining errors (both mathematical and grammatical) arc all mine.

Contents

Acronyms

CTM	Change of time method
S&P60	Standard & Poor's 60 (stock market index of 60 large Canadian companies listed on the Toronto Stock Exchange)
MRAM	Mean-reverting asset model
MRRNAM	Mean-reverting risk-neutral asset model
SDE	Stochastic differential equations
AECO	Alberta Energy Company
SABR	"Stochastic alpha, beta, rho" stochastic volatility model
LIBOR	London Interbank Offered Rate
MRSVM	Mean-reverting stochastic volatility model
S&P GSCI	Standard & Poor's Goldman Sachs Commodity Index
AIG	American International Group
DJ-AIGCI	Dow Jones-AIG Commodity Index
CBOE	Chicago Board Options Exchange
ETF	Exchange-traded fund
VIX	Volatility Index
OVX	CBOE Crude Oil ETF Volatility Index, 'Oil VIX'

Chapter 1
Introduction to the Change of Time Methods: History, Finance and Stochastic Volatility

"Both Aristotle and Newton believed in absolute time...Time was completely separate from and independent of space...However, we have had to change our ideas about space and time..."—Stephen Hawking "A Brief History of Time"

Abstract In this Chapter, we provide a historical context for the development of change of time methods (CTM) and their connections with finance in general and stochastic volatility methods in particular.

1.1 Introduction to the Change of Time Methods

Let $(\Omega, \mathscr{F}, \mathscr{F}_t, P)$ be a filtered probability space, $t \geq 0$. Something we shall frequently use in this book is the notion of: 1) Brownian motion B_t (or Wiener process W_t), a process with independent Gaussian (normal) increments and continuous trajectories [Einstein (1905) used it when analyzing the chaotic motion of particles in a liquid]; 2) the stochastic differential equation $dX_t = a(X_t, t)dt + \sigma(X_t, t)dB_t$ (describing the diffusion process X_t with drift a and diffusion σ) with local Lipschitz and linear growth conditions for the coefficients a and σ; 3) the martingale M_t (stochastic model for fair game), meaning $E|M_t| < +\infty$ and $E(M_t|\mathscr{F}_s) = M_s, s \leq t$, where M_t is a stochastic process on the filtered probability space mentioned above; 4) the Lévy process L_t (the process that contains deterministic drift, diffusion, and jumps) (i.e. a stochastically continuous process with stationary and independent increments) (see, e.g. Jacod and Shiryaev 1987 and Applebaum 2003).

The main idea of the change of time method is to find a simpler representation for a stochastic process of complicated structure by combining a simple stochastic process with a change of time. For example, if we consider a Brownian motion B_t (or Lévy process L_t) as a simple process and X_t as a complicated process that

A. Swishchuk, *Change of Time Methods in Quantitative Finance*,
SpringerBriefs in Mathematics, DOI 10.1007/978-3-319-32408-1_1

satisfies the following stochastic differential equation $dX_t = a(X_t,t)dt + \sigma(X_t,t)dB_t$ (or $dX_t = a(X_t,t)dt + \sigma(X_t,t)dL_t$) on the latter filtered probability space, then the question is: can we represent X_t in the following form:

$$X_t = B_{T_t}, \quad (or \quad X_t = L_{T_t})$$

where T_t is a change of time process? In many cases, the answer is "yes". In this book, we shall show those cases and many applications of them. In general, the procedure of change of time means that we proceed from old (physical or the calendar) time t to a new (operational or business) time t' with $t' = T_t$ in a such way to be able to construct our initial complicated process X_t (in an old time) through a simple process $\hat{X}_t$ (in a new time) that satisfies the relation $X_t = \hat{X}_{T_t}$. If we define $t = \hat{T}_{t'}$, then $\hat{X}_{t'} = X_{\hat{T}_{t'}}, t' > 0$.

1.1.1 A Brief History of the Change of Time Method

To the best of the author's knowledge, Wolfgang Doeblin (see Doeblin 2000; Lévy 1955; Lindvall 1991; and also Bru and Yor 2002 for details) was the first mathematician who introduced the change of time method into the theory of stochastic processes. He took the martingale point of view in his analysis of the paths of an inhomogeneous real-valued diffusion $X_t, t \geq 0$, starting from x, with drift coefficient $a(x,t)$ and diffusion coefficient $\sigma(x,t)$. If we define $Y_t := X_t - x - \int_0^t a(X_s,s)ds$ and $h_t := \int_0^t \sigma^2(X_s,s)ds$, then he proved first that Y_t and $Y_t^2 - h_t$ are martingales without mentioning the notion of a martingale (see Doeblin (2000)). Secondly, Doeblin introduced the time change $\theta(\tau) := \inf\{t : m_t > \tau, \quad \tau > 0\}$ and showed that $B(\tau) := Y_{\theta(\tau)}$ is a Brownian motion. In fact, he proved that there exists a Brownian motion $B(s), s \geq 0$, such that $Y_t := B(h_t)$. All in all, Doeblin has obtained the representation of X_t as

$$X_t = x + B(h_t) + \int_0^t a(X_s,s)ds. \tag{1.1}$$

Of course, at that time the general notion of a martingale did not exist (see for more details Bru and Yor 2002). The notion of a positive martingale and its denomination (which Doeblin does not use) are due to Ville, in his 1939 thesis (see Ville 1939) (martingale property was also used by Levy under another name, "condition C", since 1934). Several years later, Ito (1942a,b) (see also Ito 1951a,b) presented X_t in the form of a stochastic differential equation:

$$X_t = x + \int_0^t a(X_s,s)ds + \int_0^t \sigma(X_s,s)dB(s), \tag{1.2}$$

where $B(s)$ is a Brownian motion. If we compare Doeblin's and Ito's representations ((1.1) and (1.2)), then we can obtain

$$\int_0^t \sigma(X_s, s)ds = B(h_t). \tag{1.3}$$

The latter result (1.3) would be understood, of course, in a general setting many years later with the Dambis (1965) and Dubins and Schwartz (1965) representation of a continuous martingale M_t as

$$M_t = B(<M>_t), \tag{1.4}$$

where $B(u)$ is a Brownian motion and $<M>$ is the quadratic variation of M. The idea of associating a compensated process Y_t, which follows the trajectories of a standard Brownian motion, with a diffusion X_t, is presented in the works of Levy on additive processes (see Lévy 1934, 1935, 1937, 1948) and in the seminal paper of Kolmogorov (1931). However, as Bru and Yor (2002) mentioned, "Doeblin's method goes much further and the change of time which he adopts seems to be original. Usually it is attributed to Volkonskii (1958); see, e.g. Dynkin (1965) and Williams (1979). In any case, there does not seem to be much use of random time changes in the study of diffusion before the end of the fifties". However, we would like to mention that Bochner (1949) also used the notion of "change of time", namely, time-changed Brownian motion, before the beginning of the 1950s.

Girsanov (1960) used the change of time method to find a nontrivial weak solution to the stochastic differential equation: $dX_t = |X_t|^\alpha dB_t$, where $X_0 = 0$, B_t is a Brownian motion and $0 < \alpha < 1/2$.

It is worth mentioning here that the change of time method is closely associated with the embedding problem: to embed a process $X(t)$ in a Brownian motion is to find a Brownian motion (or Wiener process) $B(t)$ and an increasing family of stopping times T_t such that $B(T_t)$ has the same joint distribution as $X(t)$. Skorokhod (1964) first treated the embedding problem, showing that the sum of any sequence of independent random variables with mean of zero and a finite variation could be embedded in a Brownian motion using stopping times. See also Monroe (1972).

Dambis (1965) and Dubins and Schwartz (1965) independently showed that every continuous martingale could be embedded in a Brownian motion (in the sense of (1.4) above). Feller (1936, 1966) introduced a subordinated process $X(T_t)$ for a Markov process $X(t)$ with T_t as a process having independent increments. T_t was called "randomized operational time". Huff (1969) showed that every process of pathwise bounded variation could be embedded in a Brownian motion. Knight (1971) discovered a multivariate extension of the Dambis (1965) and Dubins and Schwartz (1965) result. Monroe (1972) proved that every right continuous martingale could be embedded in a Brownian motion. Meyer (1971) and Papangelou (1972) independently discovered Knight's (1971) result for point processes.

Clark (1973) introduced the change of time method into financial economics. Monroe (1978) proved that a process can be embedded in Brownian motion if and only if this process is a local semimartingale. Johnson (1979) introduced a time-changed stochastic volatility model in continuous time. Ikeda and Watanabe (1981) introduced and studied the change of time to find the solutions of stochastic differential equations. Rosinski and Woyczinski (1986) considered time changes for integrals over stable Lévy processes. Lévy processes can also be used as a time change for other Lévy processes (subordinators). Johnson and Shanno (1987) studied pricing of options using a time-changed stochastic volatility model.

Madan and Seneta (1990) introduced the variance gamma process (i.e. Brownian motion with drift time changed by a gamma process). Kallsen and Shiryaev (2002) showed that the Rosinski-Woyczynski-Kallenberg result cannot be extended to any other Lévy processes other than the symmetric α-stable processes. Kallenberg (1992) considered time change representations for stable integrals. Geman et al. (2001a) considered time changes ("business times") for Lévy processes.

Barndorff-Nielsen et al. (2002) studied the relationship between subordination and stochastic volatility models using a change of time (they called it the T_t-"chronometer"). Carr et al. (2003) used subordinated processes to construct stochastic volatility for Lévy processes (T_t being "business time").

Carr et al. (2003) also used a change of time method to introduce stochastic volatility into a Lévy model to achieve a leverage effect and a long-term skew.

Swishchuk (2004) applied a change of time method for pricing variance, volatility, covariance, and correlation swaps for Heston model. A change of time method was applied to option pricing for mean-reverting models in energy markets in Swishchuk (2008b). Pricing of Levy-based interest rate derivatives based on the change of time method was considered in Swishchuk (2008a). An overview of the change of time method in mathematical finance and its applications was presented in Swishchuk (2007). Applications of the change of time method to multifactor Levy-based model for pricing of financial and energy derivatives were considered in Swishchuk (2009).

The book Option Prices as Probabilities by Profeta et al. (2010) also relies on change of time method. The article "Time Change" by Veraart and Winkel (2010) reviews the theory on time-changed stochastic processes and relates them to stochastic volatility models in finance. There are several papers written by Linetsky (2014) and Mendoza-Arriaga et al. (2010, 2014) devoted to the change of time method (see, e.g. recent papers "Time-changed Markov processes in unified credit-equity modeling" by Mendoza-Arriaga et al. (2010), "Time-changed Ornstein-Uhlenbeck processes and their applications in commodity derivative models" by Mendoza-Arriaga et al. (2014), or "Time-changed CIR default intensities with two-sided mean-reverting jumps" by Linetsky (2014)). They all use subordinator as a time change process. However, in probability theory, the term subordinator describes a particular class of stochastic processes (i.e. nondecreasing Lévy processes) and does not include all time changes. The use of time change in our case is more general as we can see later in the book.

The recent book Change of Time and Change of Measure by Barndorff-Nielsen and Shiryaev (2010) states the main ideas and results of the stochastic theory of "change of time and change of measure".

1.1.2 The Change of Time Method and Finance

The change of time method in finance is related directly to the notion of volatility, the measure for variation of price of a financial instrument (stock, etc.) over time $t \geq 0$. Often the change of time is called an "operational time" (the term first coined by Feller (1966)) or "business time" (Carr et al. 2003). This time measures the intensity of the variations/fluctuations of the prices in the financial markets. The notion of change of time is very important in finance because many prices in the financial markets can be expressed in the form of Brownian motion with a changed time, called the operational or business time.

The role of Brownian motion in finance is also hard to underestimate: besides its important role in probability and stochastic processes (central limit theorem, functional limit theorem, and so-called time-changed Brownian motion processes), it was the main component in modelling of the dynamics of financial asset prices $S_t, t \geq 0$. We would like to mention Bachelier model (1900) $S_t = S_0 + \mu t + \sigma B_t$ and Samuelson model (1965) $\log \frac{S_t}{S_0} = \mu t + \sigma B_t$ for asset prices S_t, where B_t is a Brownian motion. Another important process in finance is Poisson process, antipode to Brownian motion, which was first used by Lundberg (1903) to model the dynamics of the capital of insurance companies.

Both Brownian motion and Poisson process are main components when constructing a more general class of processes in finance and insurance, namely, Lévy processes (see Lévy (1934, 1935, 1937, 1954); Applebaum 2003). To go further, we mention that even more general processes recently have been used to construct many financial and insurance models, namely, processes with independent increments and semimartingales (see Shiryaev 2008). The latter processes are not necessarily homogeneous as in the case of Lévy processes. As we can see later, we can obtain these kinds of processes in finance if we consider Bachelier (1900) and Samuelson (1965) models, mentioned above, in the change of time mode, namely, $S_t = S_0 + \mu T(t) + \sigma B_{T_t}$ and $\log \frac{S_t}{S_0} = \mu T_t + \sigma B_{T_t}$, respectively, where T_t is a change of time process. Another important example in finance can be constructed if we take a gamma process T_t and a Brownian motion $B(t)$, independent of each other, and then form a new process $G(t)$ such that $G(t) = \mu t + \beta T_t + B_{T_t}$. This process is called the variance gamma process (or VG process) (see Madan and Seneta 1990). To catch the leverage and clustering effects , other models of stock prices in finance can be obtained using, for example, exponential Lévy models that include both stochastic volatility and change of time and also the models based on fractional Brownian motion (see Barndorff-Nielsen and Shiryaev 2010).

Many stochastic differential equations, that practitioners use in finance, can be solved using a change of time method as well. One of such equations is the Ornstein-Uhlenbeck equation (see Ornstein and Uhlenbeck 1930) (we present this equation in general form):

$$dX_t = (a(t) - b(t)X_t)dt + c(t)dW(t),$$

where $W(t)$ is a Wiener process (or Brownian motion) and a, b, c are deterministic functions of time $t \geq 0$. The solution of this equation is

$$X_t = \exp(-\int_0^t b(s)ds)[X_0 + \int_0^t a(s)\exp(\int_0^s b(u)du)ds + \int_0^t c(s)\exp(\int_0^s b(u)du)dW_s].$$

The solution of this equation can also be presented in the following form, using a change of time method (see Swishchuk 2007; Ikeda and Watanabe 1981):

$$X_t = \exp(-\int_0^t b(s)ds)[X_0 + \int_0^t a(s)\exp(\int_0^s b(u)du)ds + B_{T_t}],$$

where $T_t = \int_0^t (c(s))\exp(\int_0^s b(u)du))^2 ds$-change of time process, and B_t is a new Brownian motion. The latter Brownian motion can be obtained from the previous one, $W(t)$, by the following formulae:

$$B_t := \int_0^{\hat{T}_t} c(s)\exp(\int_0^s b(u)du)dW(s),$$

where $\hat{T}_t = \inf\{s : T_s > t\}$, where T_t is defined above.

We shall use this approach in Chapter 2 to find the solutions of many stochastic differential equations.

1.1.3 The Change of Time Method and Stochastic Volatility

Volatility is a measure for variation of price S_t of a financial instrument over time t, $t \geq 0$. We use the symbol σ for volatility, and it corresponds to the standard deviation which quantifies the amount of variation or dispersion of a set of data values S_t. Of course, σ can be a positive constant, a positive deterministic function of time $\sigma(t)$, or a stochastic process $\sigma(t, \omega)$, $\omega \in \Omega$, e.g. that satisfies some stochastic differential equation. The model for volatility that initiated the stochastic volatility model was implied volatility model: this volatility $\sigma \equiv \sigma(K, T)$ can be derived from the Black-Scholes formula for the European call option price and demonstrates the smile effect, e.g. dependency of volatility from strike price K and maturity T (see Fouque et al. 2000). This smile effect tells us that the Black-Scholes model with a constant volatility is not adequate to statistical and probabilitical structures of observable prices S_t. Merton (1973) was the first one who replaced the constant volatility σ by a deterministic function $\sigma = \sigma(t)$, $t \geq 0$. In such models there is no smile effect across strike; however, a smile effect appears for different maturities. Another way to obtain a smile effect with nonstochastic volatility is to add one more variable to the deterministic volatility $\sigma(t)$, namely, phase one, $S : \sigma \equiv \sigma(t, S)$ (see Dupire 1994). We can go further and assume that volatility depends not only on t and S but also on all proceeding values $S_u, u \leq t$, i.e. the volatility $\sigma(t, S_u; u \leq t))$

depends on all past observed prices, or volatility depends on its own past values. The latter case will be considered in Chapter 8. Besides the smile effect, mean reversion (i.e. returning of the volatility to the mean) is another important property of stochastic volatility. That's why most of modern models of stochastic volatility are assumed that the volatility is generated by another source of randomness than initial Brownian motion B_t : for example, by some process Y_t, which correlates with B_t, and this process follows some mean-reversion process, say, Ornstein-Uhlenbeck or CIR process (see Cont and Tankov 2004).

The connection of change of time with stochastic volatility can be described by the following representation: $X_t = \hat{X}_{T(t)} = \int_0^t \sigma(s, \omega) d\tilde{X}_s$, where X_t is a given process, $\sigma(t, \omega)$ is a stochastic volatility, $\hat{X}_t$ is a simple initial process, and $T(t)$ is a change of time process. In many cases in finance, the process $\tilde{X}_t$ is a Brownian motion or Lévy process. In general, the process $\tilde{X}_t$ is a semimartingale, meaning $\tilde{X}_t = \tilde{X}_0 + A_t + M_t$, where A_t is a process of bounded variation and M_t is a local martingale (see, e.g. Barndorff-Nielsen and Shiryaev 2010).

The most typical example of this connection between change of time and stochastic volatility is the following:

Let $M_t = \int_0^t \sigma(s, \omega) dB_s, t \geq 0$, where B_s is a Brownian motion and $\sigma(s, \omega)$ is a positive process such that $\int_0^t \sigma_s^2 ds < +\infty$. Then M_t can be presented in the following way:

$$M_t = \hat{B}_{T_t},$$

where $T_t := \int \sigma_s^2 ds$, $\hat{B}_t := M_{\hat{T}_t}$, and $\hat{T}_t = \inf\{s : \int_0^s \sigma_u^2 du \geq t\}$. We note that $\hat{B}_t := M_{\hat{T}_t}$ is a Brownian motion with respect to the filtration $\hat{\mathscr{F}}_t := \mathscr{F}_{\hat{T}_t}$.

Another interesting example associated with the α-stable processes L_s^α, $0 < \alpha \leq 2$ (see Applebaum 2003):

Let $X_t = \int_0^t \sigma(s, \omega) dL_s^\alpha$, $T(t) := \int_0^t |\sigma(s, \omega)|^\alpha ds < +\infty$, and $\hat{T}_t = \inf\{s \geq 0 : T_s > t\}$. Then $\hat{L}_t^\alpha := X_{\hat{T}_t}, t \geq 0$, is an α-stable process. The proof follows from the Doob optional sampling theorem and the characteristic property of semimartingales (see Jacod and Shiryaev 1987). We note that process X_t can be represented through change of time in the following way:

$$X_t = \hat{L}_{T_t}^\alpha.$$

We shall use this approach in Chapters 6 and 7 for Lévy-based and multifactor financial models.

It is worth mentioning that the only change of time process T_t that retains the Gaussian property of the time changed Brownian motion B_{T_t} is the deterministic one.

The probability literature has demonstrated that stochastic volatility models and their time-changed Brownian motion relatives are fundamental (see Shephard 2005a,b).

1.2 Structure of the Book

Chapter 2 is devoted to the definitions and general theory of the change of time method and also many approaches, including martingale, semimartingale, and stochastic differential equations (SDEs).

Chapter 3 gives an overview on many applications of change of time method. This chapter constitutes the ultimate difference between Barndorff-Nielsen-Shiryaev's book (2010) and present book.

Chapter 4 gives yet another (among many) derivation of the Black-Scholes option pricing formula using the change of time method. In this chapter we also present a brief introduction to the option pricing theory.

Chapter 5 models and Prices variance, volatility, covariance, and correlation swaps for the classical Heston model of a stock price. We also give a numerical example based on $S\&P60$ Canada Index.

Chapter 6 introduces a new delayed Heston model for pricing of variance and volatility swaps and also for hedging volatility swaps using variance swaps. Here, we use the change of time method as well and also calibrate all the parameters based on real data. This model improves the market volatility surface fitting by 44% compared to a classical Heston model.

Chapter 7 deals with an explicit option pricing formula for a mean-reverting asset in energy markets using a change of time method. We also present here a numerical example for AECO natural gas index.

Chapter 8 introduces multifactor Lévy-based financial and energy models. A change of time method is used to price many financial and energy derivatives.

All chapters contain their own list of References.

References

Applebaum, D. *Levy Processes and Stochastic Calculus*, Cambridge University Press, 2003.

Bachelier, L. Theorie de la speculation. *Ann. Ecole Norm. Sup.* 17, (1900), 21–86.

Barndorff-Nielsen, O.E., Nicolato, E. and Shephard, N. Some recent development in stochastic volatility modeling, *Quantitative Finance*, 2, 11–23, 2002.

Barndorff-Nielsen, O. E. and Shiryaev, A. N. *Change of Time and Change of Measure*. World Scientific. Singapore, 2010, 305 p..

Bochner, S. Diffusion equation and stochastic processes. *Proc. Nat. Acad. Sci. USA*, 85, 369–370, 1949.

Bru, B. and Yor, M. Comments on the life and mathematical legacy of Wolfgang Doeblin. *Finance and Stochastics*, 6, 3–47 (2002).

Carr, P., Geman, H., Madan, D. and Yor, M. Stochastic volatility for Lévy processes. Mathematical Finance, vol. 13, No. 3 (July 2003), 345–382.

Clark, P. A subordinated stochastic process model with fixed variance for speculative prices, *Econometrica*, 41, 135–156, 1973.

Cont, R. and Tankov, P. *Financial Modeling with Jump Processes*, Chapman & Hall/CRC Fin. Math. Series, 2004.

Dambis, K.E. On the decomposition of continuous submartingales, *Theory Probabability and its Appl.*, 10, 4091–410, 1965.

Doeblin, W. Sur l'equation de Kolmogoroff. Pli cachete depose le 26 fevrier 1940, ouvert le 18 mai 2000. *C.R. Acad. Sci. Paris*, Series I, 331, 1031–1187 (2000).

Dubins, L. and Schwartz, G. On continuous martingales, *Proc. Nat. Acad. Sciences, USA*, 53, 913–916, 1965.

Dupire, B. Pricing with a smile. *Risk*, 7,1, 18–20 (1994).

Dynkin, E. B. *Markov Processes*. vol. 2. Berlin-Heidelberg-New York: Springer, 1965.

Einstein, A. Uber die von der molekularkinetischen Theorie der Warme geforderte Bewegung von in rehunden Flussigkeiten suspendierten Teilchen, *Ann. Phys.* (4) 17, 549–560 (1905).

Feller, W. Zur der stochastischen prozesse (Existenz-und Eindeutigkeitssatze). *Math. Ann.*, 113, 113–160 (1936).

Feller, W. *Introduction to Probability Theory and its Applications*, v. II, Wiley & Sons, 1966.

Fouque, J.-P., Papanicolaou, G. and Sircar, K. R. *Derivatives in Financial Markets with Stochastic Volatilities*. Springer-Verlag, 2000.

Geman, H., Madan, D. and Yor, M. Time changes for Lévy processes, Math. Finance, 11, 79–96 (2001).

Girsanov, I. On transforming a certain class of stochastic processes by absolutely continuous subtitution of measures. Theory Probab. Appl., 5(1960), 3, 285–301.

Huff, B. The loose subordination of differential processes to Brownian motion, Ann. Math. Statist., 40, 1603–1609.

Ikeda, N. Watanabe, S. Stochastic Differential Equations and Diffusion Processes. North-Holland/Kodansha Ltd., Tokyo, 1981.

Ito, K. Differential equations determining a Markoff proces.. J. Pan-Japan Math. Colloq. 1077, 1352–1400 (in Japanese) (1942a).

Ito, K. On stochastic processes (I) (Infinitely divisible laws of probability). Japan J. Math. 18, 261–301 (1942b)

Ito, K. On stochastic differential equations. *Mem. Amer. Math. Soc.*, 4 (1951a).

Ito, K. On a formula concerning stochastic differentials. *Nagoya Math. J.*, 3, 55–65 (1951b).

Jacod, J. and Shiryaev, A. *Limit Theorems for Stochastic Processes*. Springer-Verlag, 1987.

Johnson, H. Option pricing when the variance rate is changing. Working paper, University of california, Los Angeles, 1979.

Johnson, H. and Shanno, D. Option pricing when the variance is changing, J. Financial Quant. Anal., 22, 143–152 (1987).

Kallenberg, O. Some time change representations of stable integrals, via predictable transformations of local martingales. *Stochastic Processes and Their Applications*, 40 (1992), 199–223.

Kallsen, J. and Shiryaev, A. Time change representation of stochastic integrals, *Theory Probab. Appl.*, vol. 46, N. 3, 522–528, 2002.

Knight, F. A reduction of continuous, square-integrable martingales to Brownian motion, in: H. Dinges, ed., Martingales, Lecture Notes in Math. No. 190 (Springer, Berlin, 1971) pp. 19–31.

Kolmogorov A. N. Uber die analytischen methoden in der Wahrscheinlichkeitsrechnung. *Math. Ann.*, 104, 149–160 (1931).

Lévy, P. Sur les integrals dont les elements sont des variables aleatoires independentes. *Ann. J. Sc. Norm. Sup.* Pisa S. 2 (3), 337–366 (1934), 4, 217–218 (1935).

Lévy, P. Theorie de l'addition des variables aleatoires, Paris: Gauthier-Villars, 1937 (2nd edn. ibid, 1954).

Lévy, P. *Processus Stochastiques et Mouvement Brownian*, 2nd ed., Gauthier-Villars, Paris, 1948 (2nd edn. ibid, 1954).

Lévy, P. Wolfgang Doeblin (V. Doblin)(1915–1940). *Rev. Histoire Sci.* 107–115 (1955).

Lindvall, T. W. Doeblin 1915–1940. *Ann. Prob.*, 19, 929–934 (1991).

Linetsky, V. Time-Changed Ornstein-Uhlenbeck Processes and Their Applications in Commodity Derivative Models. *Mathemartical Finance*, vol. 24, issue 2, pp. 289–330, 2014.

Lundberg, F. Approximerad Framstallning Sannolikhetsfunktionen. II. Aterforrsakering av Kolletivrisker. Akad. Afhandling. (Almqvist&Wiksell, Uppsala), (1903)

Madan, D. and Seneta, E. The variance gamma (VG) model for share market returns, J. Business 63, 511–524, (1990).

Mendoza-Arriaga, R., Carr, P. and Linetsky, V. Time-changed Markov processes in unified credit-equity modeling. *Mathematical Finance*, 20 (4), pp. 527–569, 2010.

Mendoza-Arriaga, R. and Linetsky, V. Time-changed CIR default intensities with two-sided mean-reverting jumps. *Ann. Appl. Probab.*, vol. 24, N. 2, pp. 811–856, 2014.

Merton, R. Theory of rational option pricing. *Bell J. Econ. and Management Sci.*, 4, 141–183 (1973).

Meyer, P. A. Demonstration simplifiee d'un theoreme de Knight, in: Seminaire de Probabilites V, Lecture Notes in Math. No. 191 (Springer, Berlin, 1971) pp. 191–195.

Monroe, I. On embedding right continuous martingales in Brownian motion, *Ann. Math. Statist.*, 43, 1293–1311, 1972.

Monroe, I. Processes that can be embedded in Brownian motion, *The Annals of Probab.*, 6, No. 1, 42–56, 1978.

Ornstein, L. and G.Uhlenbeck. On the theory of Brownian motion. *Physical Review,* 36 (1930), 823–841.

Papangelou, F. Integrability of expected increments of point processes and a related random change of scale, *Trans. Amer. Math. Soc.*, 165 (1972) 486–506.

Profeta, B., Roynette, B. and Yor, M. *Option Prices as Probabilities*. Springer-Verlag, Berlin-Hedelberg, 2010.

Rosinski, J. and Woyczinski, W. On Ito stochastic integration with respect to p-stable motion: Inner clock, intagrability of sample paths, double and multiple integrals, *Ann. Probab.*, 14 (1986), 271–286

Samuelson, P. Rational theory of warrant pricing. *Industrial Management Rev.*, 6, (1965), 13–31.

Shephard, N. *Stochastic Volatility: Selected Readings*. Oxford: Oxford University Press, 2005a.

Shephard, N. Stochastic Volatility. *Working paper*, Oxford: Oxford University Press, 2005b.

Shiryaev, A. *Essentials of Stochastic Finance*, World Scientific, 2008.

Skorokhod, A. *Random Processes with Independent Increments*, Nauka, Moscow, 1964. (English translation: Kluwer AP, 1991).

Skorokhod, A. *Studies in the Theory of Random Processes*, Addison-Wesley, Reading, 1965.

Swishchuk, A. Modelling and valuing of variance and volatility swaps for financial markets with stochastic volatilites, *Wilmott Magazine,* Technical Article, N0. 2, September, 2004, 64–72.

Swishchuk, A. Change of time method in mathematical finance, *Canad. Appl. Math. Quart.*, vol. 15, No. 3, 2007, 299–336.

Swishchuk, A. Lévy-based interest rate derivatives: change of time and PIDEs, *Canad. Appl. Math. Quart.*, v. 16. No. 2, 2008a.

Swishchuk, A. Explicit option pricing formula for a mean-reverting asset in energy market, *J. of Numer. Appl. Math.*, Vol. 1(96), 2008b, 216–233.

Swishchuk, A. Multi-factor Lévy models for pricing of financial and energy derivatives, *Canad. Appl. Math. Quart.*, v. 17, N0. 4, 2009.

Veraart, A. and Winkel, M. Time Change. *Encyclopedia of Quantitative Finance*. Wiley, 2010.

Ville, J. Etude critique de la notion de collectif, These sci. math. Paris, (fascicule III de la Collection de monographies des probabilites, publiee sous la direction de M. Emile Borel). Paris: Gautier-Villars, 1939.

Volkonskii, V. A. Random substitution of time in strong Markov processes. *Teor. Veroyatnost. i Primenen.*, 3, 332–350 (1958).

Williams, D. *Diffusion, Markov Processes and Martingales I.* Fiundations. New York, Wiley, 1979.

Chapter 2
Change of Time Methods: Definitions and Theory

"To know, is to know that you know nothing. That is the meaning of true knowledge".—Socrates.

Abstract In this chapter, we consider the general theory of a change of time method (CTM). One of probabilistic methods which is useful in solving stochastic differential equations (SDEs) arising in finance is the "*change of time method*". We give the definition of CTM and describe CTM in martingale, semimartingale, and the SDEs settings. We also point out the association of CTM with subordinators and stochastic volatilities.

2.1 Change of Time Methods: Definitions, Properties, and Theory

2.1.1 A Change of Time Process: Definition and Properties

Let $(\Omega, \mathscr{F}, \mathscr{F}_t, P)$ be a filtered probability space with a sample space Ω, σ-algebra $\mathscr{F}$ of subsets of Ω and probability measure P. The filtration $\mathscr{F}_t, t \geq 0$, is a nondecreasing right-continuous family of sub-σ-algebras of $\mathscr{F}$.

Definition of a Change of Time Process. A *change of time process* is a right-continuous increasing, $[0, +\infty]$-valued and $\mathscr{F}_t$-adapted process $(T_t)_{t \in R_+}$ such that $\lim_{t \to +\infty} T_t = +\infty$. T_t is also a stopping time for any $t \in R_+$.

By $\hat{\mathscr{F}}_t := \mathscr{F}_{T_t}$, we define the *time-changed filtration* $(\hat{\mathscr{F}}_t)_{t \in R_+}$. The *inverse time change* $(\hat{T}_t)_{t \in R_+}$ is defined as $\hat{T}_t := \inf\{s \in R_+ : T_s > t\}$. We note that $\hat{T}_t$ is an increasing process and that $\lim_{t \to +\infty} \hat{T}_t = +\infty$. Furthermore, $\hat{T}_t$ is an $\mathscr{F}_t$-stopping time. Let X_t be an $\mathscr{F}_t$-adapted process. By this, we may define $X_{\hat{T}_t}$. Then $X_{\hat{T}_t}$ is an $\hat{\mathscr{F}}_t$-adapted process, and this process is called the *time change of X_t by T_t*.

© The Author 2016

A. Swishchuk, *Change of Time Methods in Quantitative Finance*,
SpringerBriefs in Mathematics, DOI 10.1007/978-3-319-32408-1_2

One of the examples of change of time is the following:

Let A_t be an $\mathcal{F}_t$-adapted, increasing, right-continuous random process with $A_0 = 0$. Define the following process:

$$\hat{T}_t = \inf\{s : A_s > t\}, \quad t \geq 0.$$

Then the process $\hat{T}_t$ is a change of time process. We call A_t the *process generating the change of time* $\hat{T}_t$. We note that the process T_t (see definition above) coincides with A_t in this case. It means that change of time processes T_t and $\hat{T}_t$ is a mutually inverse process—someone may construct $\hat{T}_t$ using T_t,

$$\hat{T}_t = \inf\{s : T_s > t\},$$

or may construct T_t using $\hat{T}_t$

$$T_t = \inf\{s : \hat{T}_s > t\}.$$

We also note that

$$T_{\hat{T}_t} = t \quad \text{and} \quad \hat{T}_{T_t} = t.$$

We would also like to mention the change of time in Lebesgue-Stieltjes integrals, which is well known from calculus. If we take $A_t, A_0 = 0$ as a deterministic increasing continuous function and $f(t)$ as a nonnegative Borel function on $[0, +\infty)$, we may put

$$\hat{A}_t = \inf\{s : A_s > t\},$$

and then we have

$$\int_0^{\hat{A}_a} f(t) dA_t = \int_0^a f(\hat{A}_t) dt, \quad a > 0.$$

We note that $A_t = \int\{s : \hat{A}_s > t\}$ and $A_{\hat{A}_t} = t$. The latter expression can be written in the symmetric form as well:

$$\int_0^{A_a} f(t) d\hat{A}_t = \int_0^a f(A_t) dt, \quad a > 0.$$

There are many stochastic generalizations of the last two relationships for the case of A and f being some stochastic processes. One of them is the following: let $f(t, \omega)$ be a progressively measurable nonnegative stochastic process, and let $B_t(\omega)$ be an $\mathcal{F}_t$-measurable right-continuous process with bounded variation. Then

$$\int_0^{\hat{T}_a} f(t, \omega) dB_t(\omega) = \int_0^a f(\hat{T}_t, \omega) dB_{\hat{T}_t}(\omega),$$

where $\hat{T}_t$ is the inverse change of time process. For example, if $f(t, \omega) = F(T(t))$, where $T_t = A_t$, and A_t is a continuous and strictly increasing process generating the change of time $\hat{T}_t$ (see above), then

$$\int_0^{\hat{T}_a} F(T_t)dB_t(\omega) = \int_0^a F(t)dB_{\hat{T}_t}(\omega).$$

Of course, if $B_t = t$, then

$$\int_0^{\hat{T}_a} F(T_t)dt = \int_0^a F(t)d\hat{T}_t,$$

and if $B_t = T_t$, then

$$\int_0^{\hat{T}_a} F(T_t)dT_t = \int_0^a F(t)dt.$$

See Ikeda and Watanabe (1981) and Barndorff-Nielsen and Shiryaev (2010) for more details.

2.1.2 CTM: Martingale and Semimartingale Settings

The general theory of time changes for martingale and semimartingale theories is well known (see Ikeda and Watanabe 1981). We will give a brief overview of those results.

The following result on martingales and a change of time process belongs to Dambis (1965) and Dubins and Schwartz (1965): Suppose M_t is a square integrable local continuous martingale such that $\lim_{t \to +\infty} <M>(t) = +\infty$ a.s., and define $\hat{T}_t := \inf\{u : <M>_u > t\}$ and $\hat{\mathscr{F}}_t = \mathscr{F}_{\hat{T}_t}$. Then the time-changed process $B(t) := M_{\hat{T}_t}$ is an $\hat{\mathscr{F}}_t$-Brownian motion. Also, $M_t = B(<M>_t)$. Thus, M_t can be presented by an $\hat{\mathscr{F}}_t$-Brownian motion $B(t)$ and an $\mathscr{F}_t$-stopping time $<M>_t$. Here, $< \cdot >$ defines a predictable quadratic variation. One of the examples of this result was considered in section 1.2. for a continuous local martingale $M_t = \int_0^t \sigma_s(\omega)dB(s)$, where B_t is a Brownian motion and $\sigma_t(\omega)$ is a positive process such that $\int_0^t \sigma_s^2(\omega)ds < +\infty$. In this case, $\hat{T}_t = \inf\{s : \int_0^s \sigma_u^2(\omega)du \ge t\}$ and $T_t = \int_0^t \sigma_s^2(\omega)ds$.

This result was generalized by Knight (1971) for a d-dimensional case: Let M_t^i be square integrable local continuous martingales, $i = 1, 2, ..., d$, such that $<M^i, M^j>_t = 0$ if $i \ne j$ and $\lim_{t \to +\infty} <M^i>_t = +\infty$ a.s. If $\hat{T}_t^i = \inf\{u : <M^i>_u > t\}$, then $\mathbf{B}(t) = (B^1(t), B^2(t), ..., B^d(t))$ is a d-dimensional Brownian motion, where $B^i(t) = M_{\hat{T}_t^i}^i$, $i = 1, 2, ..., d$.

One of the main properties of the semimartingale X_t with respect to the CTM is the following (see Liptser and Shiryaev 1989): If X_t is a semimartingale with respect to a filtration $\mathscr{F}_t$, then the changed time process $X_{\hat{T}_t}$ is also a semimartingale with respect to the filtration $\hat{\mathscr{F}}_t$ (see sec. 1.1).

If we have the triplet of predictable characteristics (B_t, C_t, ν) for a semimartingale X_t, then the triplet of the time-changed semimartingale $X_{\hat{T}_t}$ is determined as $(B_{\hat{T}_t}, C_{\hat{T}_t}, I_G \nu_{\hat{T}_t})$ (see Kallsen and Shiryaev 2002).

The connection of semimartingales, Brownian motions, and CTM is described by the Monroe result (see Monroe 1978): if X_t is a semimartingale, then there exists a filtered probability space with Brownian motion $\hat{B}_t$ and a change of time T_t on it such that the distribution of X_t coincides with the distribution of $\hat{B}_{T_t}$, i.e.

$$X_t =^{law} \hat{B}_{T_t}. \tag{2.1}$$

Let us now consider a counting process N_t with respect to the filtration $\mathscr{F}_t$ and with the continuous compensator A_t such that $N_t = A_t + M_t$, where M_t is a local martingale. Here, $< M >= A$. Let us then define time change as $\hat{T}_t = \inf\{s :< M >_s > t\}$. If we suppose that $< M >_{+\infty} = +\infty$, then the following process:

$$\hat{N}_t := N_{\hat{T}_t}$$

is a standard Poisson process with the intensity parameter $\lambda = 1$. We note that the initial counting process N_t can be expressed in the following way:

$$N_t = \hat{N}_{T_t},$$

where $T_t =< M >_t$. Here, we note that $M_t = \hat{M}_{T_t}$, where $\hat{M}_t = \hat{N}_t - t$ is a Poisson martingale (see Liptser and Shiryaev 2001 for more details).

Suppose that we have a nondecreasing Lévy process X_t and a Brownian motion $\hat{B}_t$ independent of X_t. Then we can find a change of time T_t such that

$$X_t = \hat{B}_{T_t} \tag{2.2}$$

holds with a probability one. This change of time T_t can be found as

$$T_t = \inf\{s : \hat{B}_s = X_t\}.$$

We mention that a semimartingale X_t can be presented in the form of (2.1) with continuous change of time T_t if and only if the process X_t is a continuous local martingale (see Huff 1969 and Cherny and Shiryaev 2002 for more details).

2.1.3 CTM: Subordinators and Stochastic Volatility

We note that if the process $\hat{T}_t$ (see sec. 1.1) is a Lévy process, then $\hat{T}_t$ is called a *subordinator*. Feller (1966) introduced a subordinated process X_{τ_t} for a Markov process X_t and τ_t a process with independent increments. τ_t was called a "randomized operational time". Increasing Lévy processes can also be used as a time change for other Lévy processes (see Applebaum 2004; Barndorff-Nielsen et al. 2001; Barndorff-Nielsen et al. 2003; Bertoin (1996); Cont and Tankov 2004; Schoutens 2003). Lévy processes of this kind are called subordinators. They are very important ingredients for building Lévy-based models in finance (see Cont and Tankov 2004; Schoutens 2003). If S_t is a subordinator, then its trajectories are almost surely increasing, and S_t can be interpreted as a "time deformation" and used to "time change" other Lévy processes. Roughly, if $(X_t)_{t\geq 0}$ is a Lévy process and $(S_t)_{t\geq 0}$ is a subordinator independent of X_t, then the process $(Y_t)_{t\geq 0}$ defined by $Y_t := X_{S_t}$ is a Lévy process (see Cont and Tankov 2004). This time scale has the financial interpretation of business time (see Geman et al. 2001), that is, the integrated rate of information arrival. Using the subordinator S_t and a Brownian motion $\hat{B}_t$ that is independent of S_t, we can construct many stochastic processes such as $X_t = \hat{B}_{S_t}$. For example, for the Cauchy process $S_t = \inf\{s : B_s > t\}$, where B_s is a standard Brownian motion independent of $\hat{B}_t$; for generalized hyperbolic Lévy processes, S_t is generated by the nonnegative infinitely divisible random variable having generalized inverse Gaussian distribution (the normal inverse Gaussian and hyperbolic Lévy processes are particular cases of the generalized hyperbolic Lévy processes).

The time change method was used to introduce stochastic volatility into a Lévy model to achieve the leverage effect and a long-term skew (see Carr et al. 2003). In the Bates (1996) model, the leverage effect and long-term skew were achieved using correlated sources of randomness in the price process and the instantaneous volatility. The sources of randomness are thus required to be Brownian motions. In the Barndorff-Nielsen et al. (2001, 2002) model, the leverage effect and long-term skew are generated using the same jumps in the price and volatility without a requirement for the sources of randomness to be Brownian motions. Another way to achieve the leverage effect and long-term skew is to make the volatility govern the time scale of the Lévy process driving jumps in the price. Carr et al. (2003) suggested the introduction of stochastic volatility into an exponential-Lévy model via a time change. The generic model here is $S_t = \exp(X_t) = \exp(Y_{v_t})$, where $v_t := \int_0^t \sigma_s^2 ds$. The volatility process should be positive and mean-reverting (i.e. an Ornstein-Uhlenbeck or Cox-Ingersoll-Ross processes). Barndorff-Nielsen et al. (2003) reviewed and placed in the context some of their recent work on stochastic volatility models including the relationship between subordination and stochastic volatility.

In general setting, the connection between stochastic volatility and change of time can be described in the following way :

Let $X_t = \int_0^t H_s dB_s$, where H_s is the adapted process such that $\int_0^t H_s^2 ds < +\infty$, $\int_0^{+\infty} H_s^2 ds = +\infty$ and B_t is a Brownian motion. Then the process $\hat{B}_t := X_{\hat{T}_t}$, where $\hat{T}_t = \inf\{s : <X>_s > t\}$, is a Brownian motion. Moreover, the process X_t has the following representation $X_t = \hat{B}_{T_t}$, where $T_t = <X>_t = \int_0^t H_s^2 ds$.

In the case of α-stable processes Y_t^α instead of Brownian motion B_t in the integral $X_t = \int_0^t H_s dY_s^\alpha$, we have a similar result for $T_t = \int_0^t |H_s|^\alpha ds < +\infty$ and $\int_0^{+\infty} |H_s|^\alpha ds = +\infty$. If we set $\hat{T}_t = \inf\{s : T_s >> t\}$, then $\hat{Y}_t^\alpha = X_{\hat{T}_t}$ is an α-stable process and $X_t = \hat{Y}_{T_t}^\alpha$.

The main difference between the change of time method and the subordinator method is that in the former case, the change of time process T_t depends on the process X_t, but in the latter case, the subordinator S_t and Lévy process X_t are independent.

2.1.4 CTM: Stochastic Differential Equations (SDEs) Setting

2.1.4.1 General Result

We consider the following generalization of the previous results to the SDE of the following form (without a drift):

$$(2.1) \qquad dX(t) = \alpha(t, X(t))dW(t),$$

where $W(t)$ is a Brownian motion and $\alpha(t, X)$ is a continuous and measurable (by t and X) function on $[0, +\infty) \times R$.

The reason to consider this equation is the following: if we solve the equation, then we can solve a more general equation with the drift $\beta(t, X)$ by *drift transformation method* or *Girsanov transformation* (see Ikeda and Watanabe 1981, Chapter 4, Section 4).

Theorem 2.1 (Ikeda and Watanabe 1981, Chapter IV, Theorem 4.3). *Let $\tilde{W}(t)$ be an one-dimensional $\mathscr{F}_t$-Wiener process with $\tilde{W}(0) = 0$, given on a probability space $(\Omega, \mathscr{F}, (\mathscr{F}_t)_{t \geq 0}, P)$ and let $X(0)$ be an $\mathscr{F}_0$-adopted random variable. Define a continuous process $V = V(t)$ by the equality*

$$(2.2) \qquad V(t) = X(0) + \tilde{W}(t).$$

Let T_t be the change of time process (see Section 2.1.1):

$$(2.3) \qquad T_t = \int_0^t \alpha^{-2}(T_s, X(0) + \tilde{W}(s))ds.$$

If

$$(2.4) \qquad X(t) := V(\hat{T}_t) = X(0) + \tilde{W}(\hat{T}_t),$$

where

$$\hat{T}_t = \int_0^t \alpha^2(s, X(0) + \tilde{W}(\hat{T}_s))ds,$$

and $\tilde{\mathscr{F}}_t := \mathscr{F}_{\hat{T}_t}$, then there exists a $\tilde{\mathscr{F}}_t$-adopted Wiener process $W = W(t)$ such that $(X(t), W(t))$ is a solution of (2.1) on the probability space $(\Omega, \mathscr{F}, \tilde{\mathscr{F}}_t, P)$. Here, $\hat{T}_t$ is the inverse process of T_t in (2.3).

Proof. of this theorem may be found in Ikeda and Watanabe (1981), Chapter IV, Theorem 4.3.

We note that in this case,

$$(2.5) \qquad M(t) := \tilde{W}(\hat{T}_t)$$

is a martingale with quadratic variation

$$(2.6) \qquad <M>(t) = \hat{T}_t = \int_0^{\hat{T}_t} \alpha^2(T_s, X) dT_s = \int_0^t \alpha(s, X)^2 ds,$$

and $\hat{T}_t$ satisfies the equation

$$(2.7) \qquad \hat{T}_t = \int_0^t \alpha^2(s, X(0) + \tilde{W}(\hat{T}_s)) ds.$$

We also remark that

$$(2.8) \qquad W(t) = \int_0^t \alpha^{-1}(s, X(s)) d\tilde{W}(\hat{T}_s) = \int_0^t \alpha^{-1}(s, X(s)) dM(s)$$

and

$$X(t) = X(0) + \int_0^t \alpha(s, X) dW(s).$$

2.1.4.2 Corollary

The solution of the following SDE

$$(2.9) \qquad dX(t) = a(X(t)) dW(t)$$

may be presented in the following form

$$X(t) = X(0) + \tilde{W}(\hat{T}_t),$$

where $a(X)$ is a continuous measurable function, $\tilde{W}(t)$ is a one-dimensional $\mathscr{F}_t$-Wiener process with $\tilde{W}(0) = 0$, given on a probability space $(\Omega, \mathscr{F}, (\mathscr{F}_t)_{t \geq 0}, P)$ and $X(0)$ is an $\mathscr{F}_0$-adapted random variable. In this case,

$$(2.10) \qquad T_t = \int_0^t a^{-2}(X(0) + \tilde{W}(s)) ds,$$

and

$$(2.11) \qquad \hat{T}_t = \int_0^t a^2(X(0) + \tilde{W}(\hat{T}_s)) ds.$$

(See Ikeda and Watanabe 1981, Chapter IV, Example 4.2).
We note that

$$M(t) := \tilde{W}(\hat{T}_t)$$

is a martingale with a quadratic variation

$$< M > (t) = \hat{T}_t = \int_0^{\hat{T}_t} a^2(X)dT_s = \int_0^t a(X)^2 ds.$$

We also remark that

$$W(t) = \int_0^t a^{-1}(X(s))d\tilde{W}(\hat{T}_s) = \int_0^t a^{-1}(X(0) + \tilde{W}(\hat{T}_s)))d\tilde{W}(\hat{T}_s)$$

and

$$X(t) = X(0) + \int_0^t a(X(s))dW(s).$$

2.1.4.3 One-Factor Diffusion Models and Their Solutions Using CTM

In this section, we introduce well-known one-factor diffusion models (used in finance) described by SDEs and driven by a Brownian motion (so-called Gaussian models).

For one-factor Gaussian models, we define the following well-known processes:

1. *The geometric Brownian motion:* $dS(t) = \mu S(t)dt + \sigma S(t)dW(t)$;
2. *The continuous-time GARCH process:* $dS(t) = \mu(b - S(t))dt + \sigma S(t)dW(t)$;
3. *The Ornstein-Uhlenbeck (1930) process:* $dS(t) = -\mu S(t)dt + \sigma dW(t)$;
4. *The Vasićek (1977) process:* $dS(t) = \mu(b - S(t))dt + \sigma dW(t)$;
5. *The Cox et al. (1985) process:* $dS(t) = k(\theta - S(t))dt + \gamma\sqrt{S(t)}dW(t)$;
6. *The Ho and Lee (1986) process:* $dS(t) = \theta(t)dt + \sigma dW(t)$;
7. *The Hull and White (1987) process:* $dS(t) = (a(t) - b(t)S(t))dt + \sigma(t)dW(t)$;
8. *The Heath et al. (1992) process:* Define the forward interest rate $f(t,s)$, for $t \leq s$, characterized by the following equality $P(t,u) = \exp[-\int_t^u f(t,s)ds]$ for any maturity u. $f(t,s)$ represents the instantaneous interest rate at time s as "anticipated" by the market at time t. It is natural to set $f(t,t) = r(t)$. The process $f(t,u)_{0 \leq t \leq u}$ satisfies an equation

$$f(t,u) = f(0,u) + \int_0^t a(v,u)dv + \int_0^t b(f(v,u))dW(v),$$

where the processes a and b are continuous. We note that the last SDE may be written in the following form: $df(t,u) = b(f(t,u))(\int_t^u b(f(t,s)))ds + b(f(t,u)) d\hat{W}(t)$, where $\hat{W}(t) = W(t) - \int_0^t q(s)ds$ and $q(t) = \int_t^u b(f(t,s))ds - \frac{a(t,u)}{b(f(t,u))}$.

We use the change of time method to get the solutions of the SDEs mentioned above.

$W(t)$ below is a standard Brownian motion, and $\hat{W}(t)$ is a $(\hat{T}_t)_{t \in R_+}$-adapted standard Brownian motion on $(\Omega, \mathscr{F}, (\mathscr{F}_t)_{t \in R_+}, P)$.

1. The geometric Brownian motion: $dS(t) = \mu S(t)dt + \sigma S(t)dW(t)$. Solution $S(t) = e^{\mu t}[S(0) + \hat{W}(\hat{T}_t)]$, where $\hat{T}_t = \sigma^2 \int_0^t [S(0) + \hat{W}(\hat{T}_s)]^2 ds$.

2. The continuous-time GARCH process: $dS(t) = \mu(b - S(t))dt + \sigma S(t)dW(t)$. Solution $S(t) = e^{-\mu t}(S(0) - b + \hat{W}(\hat{T}_t)) + b$, where $\hat{T}_t = \sigma^2 \int_0^t [S(0) - b + \hat{W}(\hat{T}_s) + e^{\mu s}b)^2 ds$.

3. The Ornstein-Uhlenbeck process: $dS(t) = -\mu S(t)dt + \sigma dW(t)$, solution $S(t) = e^{-\mu t}[S(0) + \hat{W}(\hat{T}_t)]$, where $\hat{T}_t = \sigma^2 \int_0^t (e^{\mu s}[S(0) + \hat{W}(\hat{T}_s)])^2 ds$.

4. The Vasiček process: $dS(t) = \mu(b - S(t))dt + \sigma dW(t)$, solution $S(t) = e^{-\mu t}[S(0) - b + \hat{W}(\hat{T}_t)]$, where $\hat{T}_t = \sigma^2 \int_0^t (e^{\mu s}[S(0) - b + \hat{W}(\hat{T}_s)] + b)^2 ds$.

5. The Cox-Ingersoll-Ross process: $dS^2(t) = k(\theta - S^2(t))dt + \gamma S(t)dW(t)$, solution $S^2(t) = e^{-kt}[S_0^2 - \theta^2 + \hat{W}(\hat{T}_t)] + \theta^2$, where $T_t = \gamma^{-2} \int_0^t [e^{kT_s}(S_0^2 - \theta^2 + \hat{W}(s)) + \theta^2 e^{2kT_s}]^{-1} ds$.

6. The Ho and Lee process: $dS(t) = \theta(t)dt + \sigma dW(t)$. Solution $S(t) = S(0) + \hat{W}(\sigma^2 t) + \int_0^t \theta(s)ds$.

7. The Hull and White process: $dS(t) = (a(t) - b(t)S(t))dt + \sigma(t)dW(t)$. Solution $S(t) = \exp[-\int_0^t b(s)ds][S(0) - \frac{a(s)}{b(s)} + \hat{W}(\hat{T}_t)]$, where $\hat{T}_t = \int_0^t \sigma^2(s)[S(0) - \frac{a(s)}{b(s)} + \hat{W}(\hat{T}_s) + \exp[\int_0^s b(u)du]\frac{a(s)}{b(s)}]^2 ds$.

8. The Heath, Jarrow, and Morton process: $f(t,u) = f(0,u) + \int_0^t a(v,u)dv + \int_0^t b(f(v,u))dW(v)$. Solution $f(t,u) = f(0,u) + \hat{W}(\hat{T}_t) + \int_0^t a(v,u)dv$, where $\hat{T}_t = \int_0^t b^2(f(0,u) + \hat{W}(\hat{T}_s) + \int_0^s a(v,u)dv)ds$.

References

Applebaum, D. (2004): *Lévy Processes and Stochastic Calculus*, Cambridge University Press.

Barndorff-Nielsen, O.E. and Shephard, N. (2001): Modelling by Lévy processes for financial econometrics, in *Lévy Processes-Theory and Applications*, Birkhauser.

Barndorff-Nielsen O.E., Mikosch, T. and Resnick, S. (eds.) (2001): *Lévy Processes: Theory and Applications*, Birkhauser.

Barndorff-Nielsen, O.E. and Shephard, N. (2002): Econometric analysis of realized volatility andits use in estimating stochastic volatility models, *J. R. Statistic Soc. B*, 64, pp. 253–280.

Barndorff-Nielsen, O.E., Nicolato, E. and Shephard, N. (2002): Some recent development in stochastic volatility modeling, *Quantitative Finance*, 2, 11–23.

Barndorff-Nielsen, O.E. and Shiryaev A. (2010): *Change of Time and Change of Measure*. World Scientific.

Bates, D. (1996): Jumps and stochastic volatility: the exchange rate processes implicit in Deutschemark options. *Rev. Fin. Studies*, 9, pp. 69–107.

Bertoin, J. (1996): *Lévy Processes*, Cambridge University Press.

Black, F. (1076): The pricing of commodity contarcts, *J. Financial Economics*, 3, 167–179.

Carr, P., Geman, H., Madan, D. and Yor, M. (2003): Stochastic volatility for Lévy processes, *Mathem. Finance*, 13, pp. 345–382.

P. Carr and L. Wu (2009): Variance risk premia, *Review of Financial Studies 22*, 1311–1341.

Cherny, A. and Shiryaev, A. (2002): Change of time and change of measure for Lévy processes. *Lecture Notes* 13, Aarhus University, Aarhus, 46p.

Cont, R. and Tankov, P. (2004): *Financial Modeling with Jump Processes*, Chapman & Hall/CRC Fin. Math. Series.

Cox, J., Ingersoll, J. and Ross, S. (1985): A theory of the term structure of interest rates, *Econometrica* 53, 385–407.

Dambis, K.E. (1965): On the decomposition of continuous submartingales, *Theory Probabability and its Appl.*, 10, 4091–410.

Dubins and Schwartz (1965): On continuous martingales, *Proc. Nat. Acad. Sciences, USA*, 53, 913–916.

Feller, W. (1966): *Introduction to Probability Theory and its Applications*, v. II, Wiley & Sons.

Geman, H., Madan, D. and Yor, M. (2001): Time changes for Lévy processes, *Mathem. Finance*, 11, pp. 79–96.

Heath, D., Jarrow, R. and Morton, A. (1992): Bond pricing and the term structure of the interest rates: A new methodology. *Econometrica*, 60, 1 (1992), pp. 77–105.

Ho T.S.Y. and Lee S.-B. (1986): Term structure movements and pricing interest rate contingent claim. *J. of Finance,* 41 (December 1986), pp. 1011–1029.

Huff, B. (1969): The loose subordination of differential processes to Brownian motion, *Ann. Math. Statist.*, 40, 1603–1609.

Hull, J., and White, A. (1987): The pricing of options on assets with stochastic volatilities, *J. Finance* 42, 281–300.

Ikeda, N. and Watanabe, S. (1981): *Stochastic Differential Equations and Diffusion Processes*, North-Holland/Kodansha Ltd., Tokyo.

Kallsen, J. and Shiryaev, A. (2002): Time change representation of stochastic integrals. *Theory Probab. Appl.* 46, 3, 522–528.

Knight, F. (1971): A reduction of continuous, square-integrable martingales to Brownian motion, in: H. Dinges, ed., Martingales, Lecture Notes in Math. No. 190 (Springer, Berlin) pp. 19–31.

Liptser R. and Shiryaev A. (1989): *Theory of Martingales*. Kluwer, Dordrecht.

Liptser R. and Shiryaev A. (2001): *Statistics of Random Processes*. Vol. I: General Theory; Vol. II: Applications, 2nd edn. Springer-Verlag, Berlin.

Monroe, I. (1978): Processes that can be embedded in Brownian motion, *The Annals of Probab.*, 6, No. 1, 42–56.

Ornstein, L. and Uhlenbeck, G. (1930): On the theory of Brownian motion, *Phys. Rev.*, 36, 823–841.

Schoutens, W. (2003): *Lévy Processes in Finance: Pricing Derivatives*, Wiley.

Vasicek, O. (1977): An equilibrium characterization of the term structure, *J. Financial Econometrics*, 5, 177–188.

Chapter 3
Applications of the Change of Time Methods

"It is always more easy to discover and proclaim general principles than to apply them".—Winston Churchill.

Abstract In this chapter, we give an overview on applications of change of time methods considered in this book in Chapters 4–8. These applications include yet another (among many) derivation of the Black-Scholes formula; the derivation of option pricing formula for a mean-reverting asset in energy finance; pricing of variance, volatility, covariance, and correlation swaps for the classical Heston model; pricing of variance and volatility swaps in energy markets; pricing of financial and energy derivatives with multifactor Lévy models; and pricing of variance and volatility swaps and hedging of volatility swaps for the delayed Heston model. This chapter not only describes the applications of the change of time method but also constitutes the ultimate difference between Barndorff-Nielsen-Shiryaev's book (2010) and present book.

3.1 The Black-Scholes Formula by the Change of Time Method

In the early 1970s, Black and Scholes (1973) made a major breakthrough by deriving the pricing formula for a vanilla option written on a stock. Their model and its extensions assume that the probability distribution of the underlying cash flow at any given future time is lognormal.

There are at least three proofs of their result, including PDE and martingale approaches (see Wilmott et al. 1995; Elliott and Kopp 1999).

One of the aims of this application is to give yet another (among many) derivation of the Black-Scholes result by the change of time method.

© The Author 2016
A. Swishchuk, *Change of Time Methods in Quantitative Finance*,
SpringerBriefs in Mathematics, DOI 10.1007/978-3-319-32408-1_3

3.2 Variance and Volatility Swaps by the Change of Time Method: The Heston Model

Volatility swaps are forward contracts on future-realized stock volatility, while the variance swaps are similar contract on variance, the square of the future volatility; both these instruments provide an easy way for investors to gain exposure to the future level of volatility.

A stock's volatility is the simplest measure of its riskiness or uncertainty. Formally, the volatility σ_R is the annualized standard deviation of the stock's returns during the period of interest, where the subscript "R" denotes the observed or "realized" volatility.

The easy way to trade volatility is to use volatility swaps, sometimes called realized volatility forward contracts, because they provide pure exposure to volatility (and only to volatility).

Demeterfi et al. (1999) explained the properties and the theory of both variance and volatility swaps. They derived an analytical formula for theoretical fair value in the presence of realistic volatility skews and pointed out that volatility swaps can be replicated by dynamically trading the more straightforward variance swap.

Javaheri et al. (2002) discussed the valuation and hedging of a GARCH(1,1) stochastic volatility model. They used a general and exible PDE approach to determine the first two moments of the realized variance in a continuous or discrete context. Then they approximated the expected realized volatility via a convexity adjustment.

Brockhaus and Long (2000) provided an analytical approximation for the valuation of volatility swaps and analyzed other options with volatility exposure.

Working paper by Theoret et al. (2002) presented an analytical solution for pricing of volatility swaps, proposed by Javaheri et al. (2002). They priced the volatility swaps within framework of GARCH(1,1) stochastic volatility model and applied the analytical solution to price a swap on volatility of the $S\&P60$ Canada Index (5-year historical period: 1997 – 2002).

Although option market participants speak of volatility, it is variance, or volatility squared, that has more fundamental significance (see Demeterfi et al. 1999).

Modelling and pricing of variance, volatility, covariance, and correlation swaps for the Heston model have been considered by Swishchuk (2004). In this chapter, a new probabilistic approach, a change of time method, was proposed to study variance and volatility swaps for financial markets with underlying asset and variance that follow the Heston (1993) model. We also studied covariance and correlation swaps for the financial markets. As an application, we provided a numerical example using $S\&P60$ Canada Index to price swap on the volatility.

Variance swaps for financial markets with underlying asset and stochastic volatilities with delay were modelled and priced in the paper by Swishchuk (2005). We found some analytical close forms for expectation and variance of the realized continuously sampled variance for stochastic volatility with delay both in stationary regime and in general case. The key features of the stochastic volatility model with delay are the following: (i) continuous-time analogue of the discrete-time GARCH model; (ii) mean reversion; (iii) contains the same source of randomness as a stock price; (iv) market is complete; and (v) incorporates the expectation of log return. As applications, we provided two numerical examples using S&P60 Canada Index (1998–2002) and S&P500 Index (1990–1993) to price variance swaps with delay. Variance swaps for stochastic volatility with delay is very similar to variance swaps for stochastic volatility in the Heston model (see Swishchuk 2004), but it is simpler to model and to price it.

Variance swaps for multifactor stochastic volatility models with delay have been studied by Swishchuk (2006).

Pricing of variance swaps in Markov-modulated Brownian markets was considered in Elliott and Swishchuk (2005, 2007).

One of the aims of this application is to value variance and volatility swaps for the Heston (1993) model using a change of time method.

REMARK 1.1. An extensive review of the literature on stochastic volatility is given in Shephard (2005a,b). A detailed introduction to variance and volatility swaps, including a history and market products, may be found in Carr and Madan (1998) and Demeterfi et al. (1999). The pricing of a range of volatility derivatives, including volatility and variance swaps and swaptions, is studied in Howison et al. (2004). This chapter also contains a lot of volatility models, including those with jumps. Volatility model with jumps was first considered in Naik (1993). Parameter estimation in a stochastic drift-hidden Markov model with a cap and with applications to the theory of energy finance and interest rate modelling is studied in Hernandez et al. (2005).

REMARK 1.2. The fact that stochastic volatility models, such as the Heston model and others, are able to fit skews and smiles, while simultaneously providing sensible Greeks, has made these models a popular choice in the pricing of options and swaps. Some ideas of how to calculate the Greeks for volatility contracts may be found in Howison et al. (2004).

REMARK 1.3. We note that the change of time method was used in Swishchuk and Kalemanova (2000) to study stochastic stability of interest rates with and without jumps, in Swishchuk (2004) to model and to price variance and volatility swaps for the Heston model and in Swishchuk (2005) to price European call options for commodity prices that follow mean-reverting model.

3.3 Variance, Volatility Swaps Pricing, and Hedging for Delayed Heston Model Using the Change of Time Method

The Heston model (Heston 1993) is one of the most popular stochastic volatility models in financial industry because semi-closed formulas for vanilla option prices are available, few (five) parameters need to be calibrated, and it accounts for the mean-reverting feature of the volatility.

One might be willing, in the variance diffusion, to take into account not only its current state but also its past history over some interval $[t - \tau, t]$, where $\tau > 0$ is a constant and is called the delay. Starting from the discrete-time GARCH(1,1) model (Bollerslev 1986), a first attempt was made in this direction by Kazmerchuk et al. (2005), where a non-Markov-delayed continuous-time GARCH model was proposed. We present a variance drift-adjusted version of the Heston model which leads to a significant improvement of the market volatility surface fitting by 44% (compared to Heston). The numerical example we performed with recent market data shows a significant reduction of the average absolute calibration error[1] (calibration on 12 dates ranging from 19 Sep. to 17 Oct. 2011 for the FOREX underlying EURUSD). Our model has two additional parameters compared to the Heston model, can be implemented very easily and was initially introduced for volatility derivatives pricing. The main idea behind our model is to take into account some past history of the variance process in its (risk-neutral) diffusion. Using a change of time method for continuous local martingales, we derive a closed formula for the Brockhaus&Long approximation of the volatility swap price in this model. We also consider dynamic hedging of volatility swaps using a portfolio of variance swaps.

One of the aims of this application is to get variance and volatility swap prices using a change of time method and to get a hedge ratio for volatility swaps in the delayed Heston model.

3.4 Multifactor Lévy-Based Models for Pricing of Financial and Energy Derivatives by the Change of Time Method

We also introduce one-factor and multifactor α-stable Lévy-based models to price financial and energy derivatives. These models include, in particular, as one-factor models, the Lévy-based geometric motion model, the Ornstein-Uhlenbeck (1930), the Vasicek (1977), the Cox et al. (1985), the continuous-time GARCH, the Ho and Lee (1986), the Hull and White (1987), and the Heath et al. (1992) models and, as multifactor models, various combinations of the previous models. For example, we introduce new multifactor models such as the Lévy-based Heston model, the Lévy-based SABR/LIBOR market models, and Lévy-based Schwartz-Smith and Schwartz models. Using the change of time method for SDEs driven by α-stable Lévy processes, we present the solutions of these equations in simple and compact forms. We then apply this method to price many financial and energy derivatives such as variance swaps, options, forwards, and futures.

One of the aims of this application of the change of time method is to show how to obtain the solutions of the Lévy-based SDEs arising in financial and energy markets. We use these solutions to price, in particular, swaps and options, interest rate derivatives, and forward and future contracts.

---

[1] The average absolute calibration error is defined to be the average of the absolute values of the differences between market and model implied Black & Scholes volatilities.

3.5 Mean-Reverting Asset Model by the Change of Time Method: Option Pricing Formula

Some commodity prices, like oil and gas, exhibit the mean reversion, unlike stock price. It means that they tend over time to return to some long-term mean. This mean-reverting model is a one-factor version of the two-factor model made popular in the context of energy modelling by Pilipovic (1997). Black's model (1976) and Schwartz's model (1997) have become a standard approach to the problem of pricing options on commodities. These models have the advantage of mathematical convenience, in that they give rise to closed-form solutions for some types of option (see Wilmott 2000).

Bos et al. (2002) presented a method for the evaluation of the price of a European option based on S_t using a semi-spectral method. They did not have the convenience of a closed-form solution; however, they showed that values for certain types of option may nevertheless be found extremely efficiently. They used the following partial differential equation (see, e.g. Wilmott et al. 1995):

$$C_t' + R(S,t)C_S' + \sigma^2 S^2 C_{SS}''/2 = rC$$

for option prices $C(S,t)$, where $R(S,t)$ depends only on S and t and corresponds to the drift induced by the risk-neutral measure and r is the risk-free interest rate. Simplifying this equation to the singular diffusion equation, they were able to numerically calculate the solution.

The paper Swishchuk (2008) presents an explicit expression for a European option price, $C(S,t)$, for the mean-reverting asset S_t, using a change of time method under both physical and risk-neutral measures.

We note that the recent book by Geman (2005) covers hard and soft commodities (energy, agriculture, and metals) and analysis, economic and geopolitical issues in commodity markets, commodity price and volume risk, stochastic modelling of commodity spot prices and forward curves, real options valuation, and hedging of physical assets in the energy industry.

One of the aims of this application is to obtain an explicit expression for a European call option price on the mean-reverting model of a commodity asset using a change of time method. As we can see, if the mean-reverting level equals zero, then the option pricing formula coincides with the Black-Scholes result.

3.6 Variance and Volatility Swaps by the Change of Time Method: Energy Markets

One of the applications of CTM is devoted to the pricing of variance and volatility swaps in energy markets. We found an explicit variance swap formula and a closed-form volatility swap formula (using change of time) for an energy asset with stochastic volatility that follows a continuous-time mean-reverting GARCH (1,1) model. A numerical example is presented for AECO Natural Gas Index (1 May 1998–30 April 1999). Variance swaps are quite common in commodity, e.g. in energy market, and they are commonly traded. We consider the Ornstein-Uhlenbeck process for a commodity asset with stochastic volatility following the continuous-time GARCH model (or, one may say, Pilipovic (1998) one-factor model). The classical stochastic process for the spot dynamics of commodity prices is given by the Schwartz' model (1997). It is defined as the exponent of an Ornstein-Uhlenbeck (OU) process and has become the standard model for energy prices possessing mean-reverting features.

In our chapter Swishchuk (2013), we considered a risky asset in an energy market with stochastic volatility following a mean-reverting stochastic process satisfying the following SDE (continuous-time GARCH(1,1) model):

$$d\sigma^2(t) = a(L - \sigma^2(t))dt + \gamma\sigma^2(t)dW_t,$$

where a is a speed of mean reversion, L is the mean reverting level (or equilibrium level), γ is the volatility of volatility $\sigma(t)$ and W_t is a standard Wiener process. Using a change of time method, we find an explicit solution of this equation, and using this solution, we are able to find the variance and volatility swaps pricing formula under physical measure. Then, using the same argument, we find the option pricing formula under risk-neutral measure. We applied Brockhaus and Long (2000) approximation to find the value of a volatility swap. A numerical example for the AECO Natural Gas Index for the period 1 May 1998 to 30 April 1999 is presented.

Commodities are emerging as an asset class in their own. The range of products offered to investors range from exchange-traded funds (ETFs) to sophisticated products including principal-protected structured notes on individual commodities or baskets of commodities and commodity range-accrual or variance swap. More and more institutional investors are including commodities in their asset allocation mix, and hedge funds are also increasingly active players in commodities. For example, Amaranth Advisors lost USD 6 billion dollars during September 2006 from trading natural gas futures contracts, leading to the fund's demise. Concurrent with these developments, a number of recent papers have examined the risk and return characteristics of investments in individual commodity futures or commodity indices composed of baskets of commodity futures. However, since all but the most plain-vanilla investments contain an exposure to volatility, it is equally important for investors to understand the risk and return characteristics of commodity volatilities.

The focus on energy commodities derives from two reasons: 1) Energy is the most important commodity sector, and crude oil and natural gas constitute the largest components of the two most widely tracked commodity indices: the Standard & Poors Goldman Sachs Commodity Index (S&P GSCI) and the Dow Jones-AIG Commodity Index (DJ-AIGCI). 2) Existence of a liquid option markets: crude oil and natural gas indeed have the deepest and most liquid option markets among all commodities. The idea is to use variance (or volatility) swaps on futures contracts. At maturity, a variance swap pays off the difference between the realized variance of the future contract over the life of the swap and the fixed variance swap rate.

Since a variance swap has zero net market value at initiation, absence of arbitrage implies that the fixed variance swap rate equals the conditional risk-neutral expectation of the realized variance over the life of the swap. Therefore, e.g. the time-series average of the payoff and/or excess return on a variance swap is a measure of the variance risk premium.

Variance risk premia in energy commodities, crude oil and natural gas, has been considered by Trolle and Schwartz (2010). The same methodology as in Trolle and Schwartz (2010) was used by Carr and Wu (2009) in their study of equity variance risk premia. The idea was to use variance swaps on futures contracts. The study in Trolle and Schwartz (2010) is based on daily data from 2 January 1996 until 30 November 2006—a total of 2750 business days. The source of the data is NYMEX. Trolle and Schwartz (2010) found that: 1) The average variance risk premia are negative for both energy commodities but more strongly statistically significant for crude oil than for natural gas. 2) The natural gas variance risk premium (defined in dollar terms or in return terms) is higher during the cold months of the year. 3) Energy risk premia in dollar terms are time-varying and correlated with the level of the variance swap rate. In contrast, energy variance risk premia in return terms, particularly in the case of natural gas, are much less correlated with the variance swap rate.

The *S&P GSCI* is comprised of 24 commodities with the weight of each commodity determined by their relative levels of world production over the past five years. The DJ-AIGCI is comprised of 19 commodities with the weight of each component determined by liquidity and world production values, with liquidity being the dominant factor. Crude oil and natural gas are the largest components in both indices. In 2007, their weight were 51.30% and 6.71%, respectively, in the S&P GSCI and 13.88% and 11.03%, respectively, in the DJ-AIGCI. The Chicago Board Options Exchange (CBOE) recently introduced a Crude Oil Volatility Index (ticker symbol OVX). This index also measures the conditional risk-neutral expectation of crude oil variance, but is computed from a cross section of listed options on the United States Oil Fund (USO), which tracks the price of WTI as closely as possible. The CBOE Crude Oil ETF Volatility Index ("Oil VIX", Ticker—OVX) measures the market's expectation of a 30-day volatility of crude oil prices by applying the VIX methodology to United States Oil Fund, LP (Ticker—USO) options spanning a wide range of strike prices. We have to notice that crude oil and natural gas trade in units of 1,000 barrels and 10,000 British thermal units (mmBtu), respectively. Prices are quoted as US dollars and cents per barrel or mmBtu. The continuous-time GARCH model has also been exploited by Javaheri et al. (2002) to calculate the volatility swap for *S&P500* index. They used the PDE approach and mentioned (page 8, sec. 3.3) that 'it would be interesting to use an alternative method to calculate $F(v,t)$ and the other quantities above'. This chapter contains the exact alternative method, namely, a "change of time method", to get the variance and volatility swaps. The first paper on pricing of commodity contracts was published by Black (1976).

One of the applications of CTM is variance and volatility swap pricing in energy markets. We did not included this topic in the book due to the lack of space. We refer to Swishchuk (2013) paper for more details.

References

Applebaum, D. (2004): *Lévy Processes and Stochastic Calculus*, Cambridge University Press.

Black, F. and Scholes, M. (1973): The pricing of options and corporate liabilities, *J. Political Economy* 81, 637–54.

Black, F. (1976):The pricing of commodity contacts, *J. Financial Economics*, 3, 167–179.

Bollerslev, T. (1986): Generalized autoregressive conditional heteroscedasticity. *Journal of Economics*, 31: 307–27.

Bos, L.P., Ware, A. F. and Pavlov, B. S. (2002): On a semi-spectral method for pricing an option on a mean-reverting asset, *Quantitative Finance*, Volume 2, 337–345.

Brockhaus, O. and Long, D. (2000): Volatility swaps made simple, *RISK*, January, 92–96.

Carr, P. and Madan, D. (1998): Towards a Theory of Volatility Trading. In the book: *Volatility*, Risk book publications, http://www.math.nyu.edu/research/carrp/papers/.

Carr, P. and Wu, L. (2009): Variance risk premia, *Review of Financial Studies 22*, 1311–1341.

Cox, J., Ingersoll, J. and Ross, S. (1985): A theory of the term structure of interest rates, *Econometrica* 53, 385–407.

Demeterfi, K., Derman, E., Kamal, M. and Zou, J. (1999): A guide to volatility and variance swaps, *The Journal of Derivatives*, Summer, 9–32.

Elliott, R. and Kopp, P. (1999): *Mathematics of Financial Markets*, Springer-Verlag, New York.

Elliott, R. and Swishchuk, A. (2005): Pricing options and volatility swaps in Markov-modulated Brownian and Fractional Brownian markets, *RJE 2005 Conference*, Calgary, AB, Canada, July 24-27, 2005, 35p.

Elliott, R. and Swishchuk, A. (2007): Pricing options and variance swaps in Markov-modulated Brownian markets, In: *Hidden Markov Models in Finance*, Springer, International Series in Operations Research and Management Science, Eds.: Elliott, R. and Mamon, R.

Geman, H., Madan, D. and Yor, M. (2001): Time changes for Lévy processes, *Mathem. Finance*, 11, pp. 79–96.

Geman, H. (2005): *Commodities and Commodity Derivatives: Modelling and Pricing for Agricultural, Metals and Energy*, Wiley.

D. Heath, R. Jarrow and A. Morton: Bond pricing and the term structure of the interest rates: A new methodology. *Econometrica*, 60, 1 (1992), pp. 77–105.

Hernandez, J., Sounders, D. and Seco, L. (2005): *Parameter Estimations in a Stochastic Drift Hidden Markov Model with a Cap*, submitted to SIAM.

Heston, S. (1993): A closed-form solution for options with stochastic volatility with applications to bond and currency options, *Review of Financial Studies*, 6, 327–343

Ho T.S.Y. and Lee S.-B.: Term structure movements and pricing interest rate contingent claim. *J. of Finance*, 41 (December 1986), pp. 1011–1029.

Hobson, D. and Rogers, L. (1998): Complete models with stochastic volatility, *Math. Finance* 8, no.1, 27–48.

Howison, S., Rafailidis, A. and Rasmussen, H. (2004): *On the Pricing and Hedging of Volatility Derivatives*, Applied Math. Finance J., p.1–31.

Hull, J., and White, A. (1987): The pricing of options on assets with stochastic volatilities, *J. Finance* 42, 281–300.

Kazmerchuk, Y., Swishchuk, A. and Wu, J. (2005): A continuous-time GARCH model for stochastic volatility with delay. *Canadian Applied Mathematics Quarterly*, 13, 2: 123–149.

Javaheri, A., Wilmott, P. and Haug, E. (2002): GARCH and volatility swaps, *Wilmott Technical Article*, January, 17p.

Naik, V. (1993): Option Valuation and Hedging Strategies with Jumps in the Volatility of Asset Returns, *Journal of Finance*, **48**, 1969–84.

Ornstein L and Uhlenbeck G. (1930): On the theory of Brownian motion, *Phys. Rev.* 36, 823–841.

Pilipović, D. (1997): *Valuing and Managing Energy Derivatives*, New York, McGraw-Hill.

Shephard, N. (2005a): *Stochastic Volatility: Selected Readings*. Oxford: Oxford University Press.

Shephard, N. (2005b): *Stochastic Volatility*. Working paper, University of Oxford, Oxford.

Schwartz, E. (1997): The stochastic behaviour of commodity prices: implications for pricing and hedging, *J. Finance*, 52, 923–973.

Swishchuk, A. and Kalemanova, A. (2000): Stochastic stability of interest rates with jumps. *Theory probab. & Mathem statist.*, TBiMC Sci. Publ., v.61, Kiev, Ukraine.

Swishchuk, A. (2004): Modeling and valuing of variance and volatility swaps for financial markets with stochastic volatilities, *Wilmott Magazine*, Technical Article No2, September Issue, 64–72.

Swishchuk, A. (2005: Modelling and Pricing of Variance Swaps for Stochastic Volatilities with Delay, *Wilmott Magazine*, Technical Article, September Issue (to appear).

Swishchuk, A. (2008): Explicit option pricing formula for mean-reverting model, *J. Numer. Appl. Math.*, Vol. 1(96), 2008, pp.216–233.

Swishchuk, A. (2006): Modelling and pricing of variance swaps for multi-factor stochastic volatilities with delay, *Canadian Applied Math. Quarterly*, 14, No. 4, Winter.

Swishchuk, A. (2013): Variance and volatility swaps in energy markets. *J. Energy Markets*, v. 6, N.1, Spring.

Theoret, R., Zabre, L. and Rostan, P. (2002): Pricing volatility swaps: empirical testing with Canadian data. *Working paper*, Centre de Recherche en Gestion, Document 17-2002, July 2002.

Trolle, A. and Schwartz, E. (2010): Variance risk premia in energy commodities, *J. of Derivatives*, Spring, v. 17, No. 3, 15–32.

Vasicek, O. (1977): An equilibrium characterization of the term structure, *J. Finan. Econom.*, 5, 177–188.

Wilmott, P., Howison, S. and Dewynne, J. (1995): *The Mathematics of Financial Derivatives*, Cambridge, Cambridge University Press.

Wilmott, P. (2000): *Paul Wilmott on Quantitative Finance*, New York, Wiley.

Chapter 4
Change of Time Method (CTM) and Black-Scholes Formula

"It is said that there is no such thing as a free lunch. But the universe is the ultimate free lunch".—Alan Guth (MIT).

Abstract In this Chapter, we consider applications of the CTM to (yet one more time) obtain the well-known Black-Scholes formula for European call options. In the early 1970s, Black-Scholes (1973) made a major breakthrough by deriving a pricing formula for a vanilla option written on the stock. Their model and its extensions assume that the probability distribution of the underlying cash flow at any given future time is lognormal. We mention that there are many proofs of this result, including PDE and martingale approaches, (see Wilmott et al. 1995; Elliott and Kopp 1999). The present approach, using change of time of getting the Black-Scholes formula, was first shown in Swishchuk (2007).

4.1 A Brief Introduction to Option Pricing Theory

We use the term "asset" to describe any financial object whose value is known at present time but is liable to change in the future. Some examples include shares of a company, commodities (oil, electricity, gas, gold, etc.), currencies, etc. Now we give the definitions of options.

Definition 1 (European Call Option). A European call option gives its holder the right (but not the obligation) to purchase a prescribed asset from the writer for a prescribed price at a prescribed time in the future.

Definition 2 (European Put Option). A European call option gives its holder the right (but not the obligation) to sell a prescribed asset to the writer for a prescribed price at a prescribed time in the future.

© The Author 2016
A. Swishchuk, *Change of Time Methods in Quantitative Finance*,
SpringerBriefs in Mathematics, DOI 10.1007/978-3-319-32408-1_4

The prescribed purchase price is known as the exercise price or strike price, and the prescribed time in the future is known as the expiry date or maturity. The key question in option pricing theory is: how much should the holder pay for the privilege of holding an option? Or, how do we compute a fair option price or value?

Let $(\Omega, \mathscr{F}, \mathscr{F}_t, P)$ be a filtered probability space, $t \in [0, T]$, and $\mathscr{F}_T = \mathscr{F}$. We denote by T the expiry date, by K the strike price, and by $S(t)$ the asset price at time $t \geq 0$. Then, e.g. $S(T)$ is the asset price at the expiry date, which is not known (uncertain or random) at the time when the option is taken out. If $S(T) > K$ at expiry T, then the holder of a European call option may buy the asset for K and sell it in the market for $S(T)$, gaining an amount $S(T) - K$. If $K \geq S(T)$, on the other hand, then the holder gains nothing or zero. Therefore, the value or price of the European call option at the expiry date, denoted by $C(T)$, is

$$f_C(T) = \max\{S(T) - K, 0\}.$$

As for a European put option, the situation is the opposite. If $S(T) < K$ at expiry T, then the holder of a European put option may buy the asset for $S(T)$ in the market and sell it in the market for K, gaining an amount $K - S(T)$. If $K \leq S(T)$, on the other hand, then the holder gains nothing or zero. Therefore, the value or price of the European put option at the expiry date, denoted by $P(T)$, is

$$f_P(T) = \max\{K - S(T), 0\}.$$

We call $f_C(T)$ and $f_P(T)$ the payoff functions. The shapes of the corresponding payoff diagrams look like (ice) hockey sticks with the kinks at K.

European call and put options are the simplest and classical examples of so-called financial derivatives, meaning those derivatives indicate that their values are derived from the underlying asset (do not mix up this term with the mathematical meaning of a derivative!). There are many other financial derivatives, such as forwards, futures, swaps, etc.

The next question in the option pricing theory is: how to determine a fair value or price of the option at time $t = 0$? We denote this value or price (for European call option) by $C(0)$. To answer this question, we introduce two key concepts: discounting for interest and the no arbitrage principle (sometimes referred to as no free lunch opportunity).

Discounting interest: If we have some money in a risk-free savings account (bank deposit) and this investment grows accordingly to a continuously compounded interest rate $r > 0$, then its value increases by a factor e^{rt} over a time length t. We will use r to denote the annual rate (so that time is measured in years). The simplest example is a risk-free bank account with the amount of money $B(t)$ at time $t > 0$. If the initial deposit is $B(0)$, then at time $t > 0$ it will be $B(t) = e^{rt}B(0)$. Hence, $B(t)$ satisfies the equation $dB(t) = rB(t)dt$, $B(0) > 0$, $r > 0$, $t \geq 0$. Suppose that we have an amount $C(0)$ at time zero; then it is worth $C(t) = e^{rt}C(0)$ at time t or $C(T) = e^{rT}C(0)$ at expiry T. It means that to have $C(T)$ amount of money on saving account at time T, we have to have $C(0) = e^{-rT}C(T)$ amount of money at time $t = 0$.

No arbitrage principle: This principle means that there is never an opportunity to make a risk-free profit that gives a greater return than that provided by the interest from the bank deposit. Arguments based on the no arbitrage principle are the main tools of financial mathematics.

The key role for criteria of no arbitrage plays the problem of change of measures which is crucial in mathematical finance. We call this measure "martingale measure" or "risk-neutral measure" and denote it by Q to make it different from the initial or physical probability measure P. The technic of change of measure is based on the construction of a new probability measure Q equivalent to the given measure P and such that a process $\tilde{S}(t)$, built on the initial process $S(t)$, satisfies some "fairness" condition. In the case of mathematical finance, this process $\tilde{S}(t) = e^{-rt}S(t)$ is a martingale with respect to the new measure Q. As long as the asset value $S(t)$ at time t is random or unknown, we have to calculate the expected value of this asset. It means that expectation should be taken with respect to this martingale measure Q. We denote this expectation by E_Q and compare with E_P-expectation with respect to the initial measure P. Thus, martingale property of $\tilde{S}(t) = e^{-rt}S(t)$ means that $E_Q[e^{-rt}S(t)] = S(0)$.

The simplest explanation of using the measure Q instead of P is the following: the value of $E_Q S(t) = S(0)e^{rt}$ at time t is similar to the value of the riskless asset $B(t)$ at time t, namely, $B(t) = B(0)e^{rt}$, where $S(0)$ and $B(0)$ are initial values of stock (risky asset) and bond (riskless asset), respectively. However, it is not the case for the value of $E_P S(t)$, meaning, $E_P S(t) \neq S(0)e^{rt}$. The risk-neutral valuation means that we act on the stock market as if we put our risky asset $S(t)$ in a bank with interest rate $r > 0$. Note that $B(t)$ is a deterministic function and $S(t)$ is a stochastic function of time.

Returning to our payoff functions, it means that we have to calculate this expected value with respect to the risk-neutral measure Q : $E_Q[\max\{S(T) - K, 0\}]$.

Therefore, if the initial (at time $t = 0$) fair price of option is $C(0)$, then the value $E_Q[\max\{S(T) - K\}]$ is equal to $e^{rT}C(0)$ or

$$C(0) = e^{-rT}E_Q[\max\{S(T) - K, 0\}]. \qquad (*)$$

This formula gives the answer to our previous question: the fair price of the option at time $t = 0$ is defined by the last formula.

The situation for the European put option is similar, taking into account the payoff function $P(T) = \max\{K - S(T), 0\}$ in this case. Therefore, the fair price $P(0)$ of the European put option at time $t = 0$ is

$$P(0) = e^{-rT}E_Q[\max\{K - S(T), 0\}].$$

We note that fair prices of European call and put options satisfy the following so-called call-put parity:

$$C(0) + Ke^{-rT} = P(0) + S(0),$$

where $S(0)$ is the initial (at time $t = 0$) asset price. See Elliott and Kopp (1999) or Wilmott et al. (1995) for more details on option pricing.

4.2 Black-Scholes Formula

The well-known Black-Scholes (1973) formula states that if we have a (B,S)-security market consisting of a riskless asset $B(t)$ with a constant continuously compounded interest rate r

$$dB(t) = rB(t)dt, \quad B(0) > 0, \quad r > 0. \tag{4.1}$$

The risky asset (stock) $S(t)$ is assumed to have dynamics

$$dS(t) = \mu S(t)dt + \sigma S(t)dW(t), \quad S(0) > 0. \tag{4.2}$$

Here: $\mu \in R$ is the appreciation rate and $\sigma > 0$ is the volatility. Then the price for a European call option with the payoff function $f(T) = \max(S(T) - K, 0)$ ($K > 0$ is the strike price) has the following form:

$$C(T) = S(0)\Phi(y_+) - e^{-rT}K\Phi(y_-), \tag{4.3}$$

where

$$y_\pm := \frac{\ln(\frac{S(0)}{K}) + (r \pm \frac{\sigma^2}{2})T}{\sigma\sqrt{T}} \tag{4.4}$$

and

$$\Phi(y) := \frac{1}{\sqrt{2\pi}} \int_{-\infty}^{y} e^{-\frac{x^2}{2}} dx. \tag{4.5}$$

4.3 The Solution of SDE for a Geometric Brownian Motion Using a Change of Time Method

Lemma 1. The solution of the equation (4.2) has the following form:

$$S(t) = e^{\mu t}(S(0) + \tilde{W}(\hat{T}_t)), \tag{4.6}$$

where $\tilde{W}(t)$ is a one-dimensional Wiener process,

$$\hat{T}_t = \sigma^2 \int_0^t [S(0) + \tilde{W}(\hat{T}_s)]^2 ds$$

and

$$T_t = \sigma^{-2} \int_0^t [S(0) + \tilde{W}(s)]^{-2} ds.$$

Proof.

Set

$$V(t) = e^{-\mu t} S(t), \tag{4.7}$$

where $S(t)$ is defined in (4.2).

Applying Itô's formula to $V(t)$, we obtain

$$dV(t) = \sigma V(t) dW(t). \tag{4.8}$$

Equation (4.8) is similar to equation (2.1) of Chapter 2, with

$$a(X) = \sigma X.$$

Therefore, the solution of equation (4.8) using the change of time method (see Corollary, Section: CTM: SDEs Setting, Chapter 2) is (see also (2.4) in Chapter 2)

$$V(t) = S(0) + \tilde{W}(\hat{T}_t), \tag{4.9}$$

where $\tilde{W}(t)$ is a one-dimensional Wiener process,

$$\hat{T}_t = \sigma^2 \int_0^t [S(0) + \tilde{W}(\hat{T}_s)]^2 ds$$

and

$$T_t = \sigma^{-2} \int_0^t [S(0) + \tilde{W}(s)]^{-2} ds.$$

From (4.7) and (4.9) follows the solution of Equation (4.2) (which has the representation (4.6)).

4.4 Properties of the Process $\tilde{W}(\hat{T}_t)$

Lemma 2. Process $\tilde{W}(\hat{T}_t)$ is a mean-zero martingale with the quadratic variation

$$< \tilde{W}(\hat{T}_t) > = \hat{T}_t = \sigma^2 \int_0^t [S(0) + \tilde{W}(\hat{T}_s)]^2 ds$$

and has the following representation:

$$\tilde{W}(\hat{T}_t) = S(0)(e^{\sigma W(t) - \frac{\sigma^2}{2}t} - 1). \tag{4.10}$$

Proof. From Corollary, Section CTM: SDEs Setting, Chapter 2, follows that $\tilde{W}(\hat{T}_t)$ is a martingale with the quadratic variation

$$< \tilde{W}(\hat{T}_t) > = \hat{T}_t = \sigma^2 \int_0^t [S(0) + \tilde{W}(\hat{T}_s)]^2 ds.$$

and the process $W(t)$ has the following look:

$$W(t) = \sigma^{-1} \int_0^t [S(0) + \tilde{W}(\hat{T}_s))]^{-1} d\tilde{W}(\hat{T}_s) \tag{4.11}$$

From (4.11) we obtain the following SDE for $\tilde{W}(\hat{T}_s)$:

$$d\tilde{W}(\hat{T}_s) = \sigma[S(0) + \tilde{W}(\hat{T}_s)]dW(t).$$

Solving this equation we have the explicit expression (3.10) for $\tilde{W}(\hat{T}_s)$:

$$\tilde{W}(\hat{T}_s) = S(0)(e^{\sigma W(t) - \frac{\sigma^2}{2}t} - 1).$$

Q.E.D

We note that $E\tilde{W}(\hat{T}_s) = 0$ and $E[\tilde{W}(\hat{T}_s)]^2 = S^2(0)(e^{\sigma^2 t} - 1)$, where $E := E_P$ is an expectation under physical measure P.

Since

$$E[e^{\sigma W(t) - \frac{\sigma^2}{2}t}]^n = e^{\frac{\sigma^2}{2}n(n-1)t}, \tag{4.12}$$

we can obtain all the moments for the process $\tilde{W}(\phi_s^{-1})$:

$$E[\tilde{W}(\phi_t^{-1})]^n = S^n(0) \sum_{k=0}^n C_n^k e^{\frac{\sigma^2 t}{2}k(k-1)} (-1)^{n-k}, \tag{4.13}$$

where $C_n^k := \frac{n!}{k!(n-k)!}$, $n! := 1 \times 2 \times 3 \ldots \times n$.

Corollary 1. From Lemma 2 (see (4.6), (4.10), and (4.12)) follows that we can also obtain all the moments for the asset price $S(t)$ in (4.9), since

$$\begin{aligned} E[S(t)]^n &= e^{n\mu t} E[S(0) + \tilde{W}(\hat{T}_t)]^n \\ &= e^{n\mu t} S^n(0) E[e^{\sigma W(t) - \frac{\sigma^2 t}{2}}]^n \\ &= e^{n\mu t} S^n(0) e^{\frac{\sigma^2}{2}n(n-1)t}. \end{aligned} \tag{4.14}$$

For example, the variance of $S(t)$ is going to be

$$VarS(t) = ES^2(t) - (ES(t))^2 = S^2(0)e^{2\mu t}(e^{\sigma^2 t} - 1),$$

where $ES(t) = S(0)e^{\mu t}$ (see (4.9)).

4.5 The Black-Scholes Formula by the Change of Time Method

In a risk-neutral world, the dynamic of stock price $S(t)$ has the following look:

$$dS(t) = rS(t)dt + \sigma S(t)dW^*(t), \tag{4.15}$$

where

$$W^*(t) := W(t) + \frac{\mu - r}{\sigma}. \tag{4.16}$$

Following Section 3.2, from (4.6) we have the solution of the Equation (4.15):

$$S(t) = e^{rt}[S(0) + \tilde{W}^*(\hat{T}_t)], \tag{4.17}$$

where

$$\tilde{W}^*(\hat{T}_t) = S(0)(e^{\sigma W^*(t) - \frac{\sigma^2 t}{2}} - 1) \tag{4.18}$$

and $W^*(t)$ is defined in (4.16).

Let E_Q be an expectation under risk-neutral measure (or martingale measure) Q (i.e. process $e^{-rT}S(t)$ is a martingale under the measure Q).

Then the option pricing formula for a European call option with the payoff function

$$f_C(T) = \max[S(T) - K, 0]$$

has the following look:

$$C(T) = e^{-rT} E_Q[f(T)] = e^{-rT} E_{P^*}[\max(S(T) - K, 0)]. \tag{4.19}$$

Proposition 3.1.

$$C(T) = S(0)\Phi(y_+) - Ke^{-rT}\Phi(y_-), \tag{4.20}$$

where $y_\pm$ and $\Phi(y)$ are defined in (4.4) and (4.5).

Proof. Using a change of time method, we have the following representation for the process $S(t)$ (see (4.17)):

$$S(t) = e^{rt}[S(0) + \tilde{W}^*(\hat{T}_t)],$$

where $\tilde{W}^*(\hat{T}_t)$ is defined in (4.18). From (4.17)–(4.19), after substituting $\tilde{W}^*(\hat{T}_t)$ into (4.17) and $S(T)$ into (4.19), we get taking into account (∗) (see Section,1.1)

$$
\begin{aligned}
C(T) &= e^{-rT} E_Q[\max(S(T) - K, 0)] \\
&= e^{-rT} E_Q[\max(e^{rt}[S(0) + \tilde{W}^*(\hat{T}_t)] - K, 0)] \\
&= e^{-rT} E_Q[\max(e^{rt} S(0) e^{\sigma W^*(T) - \frac{\sigma^2 T}{2}} - K, 0)] \\
&= e^{-rT} E_Q[\max(S(0) e^{\sigma W^*(T) + (r - \frac{\sigma^2}{2})T} - K, 0)] \\
&= e^{-rT} \frac{1}{\sqrt{2\pi}} \int_{-\infty}^{+\infty} \max[S(0) e^{\sigma u \sqrt{T} + (r - \frac{\sigma^2}{2})T} - K, 0] e^{-\frac{u^2}{2}} du.
\end{aligned} \tag{4.21}
$$

Let y_0 be a solution of the following equation:

$$S(0) e^{\sigma y \sqrt{T} + (r - \sigma^2/2)T} = K,$$

namely,

$$y_0 = \frac{\ln(\frac{K}{S(0)}) - (r - \sigma^2/2)T}{\sigma\sqrt{T}}.$$

Then (4.21) may be presented in the following form:

$$C(T) = e^{-rT} \frac{1}{\sqrt{2\pi}} \int_{y_0}^{+\infty} (S(0)e^{\sigma u\sqrt{T}+(r-\frac{\sigma^2}{2})T} - K)e^{-\frac{u^2}{2}} du. \qquad (4.22)$$

Finally, straightforward calculation of the integral on the right-hand side of (4.22) gives us the Black-Scholes result:

$$\begin{aligned}
C(T) &= \tfrac{1}{\sqrt{2\pi}} \int_{y_0}^{+\infty} S(0)e^{\sigma u\sqrt{T}-\frac{\sigma^2 T}{2}} e^{-u^2/2} du - Ke^{-rT}[1 - \Phi(y_0)] \\
&= \tfrac{S(0)}{\sqrt{2\pi}} \int_{y_0-\sigma\sqrt{T}}^{+\infty} e^{-u^2/2} du - Ke^{-rT}[1 - \Phi(y_0)] \\
&= S(0)[1 - \Phi(y_0 - \sigma\sqrt{T})] - Ke^{-rT}[1 - \Phi(y_0)] \\
&= S(0)\Phi(y_+) - Ke^{-rT}\Phi(y_-),
\end{aligned}$$

where $y_\pm$ and $\Phi(y)$ are defined in (4.4) and (4.5). Q.E.D.

References

Black, F. and Scholes, M. (1973): The pricing of options and corporate liabilities, *J. Political Economy* 81, 637-54.

Elliott, R. and Kopp, P. (1999): *Mathematics of Financial Markets*, Springer-Verlag, New York.

Swishchuk, A. (2007): Change of time method in mathematical finance. *CAMQ*, v. 15, No. 3.

Wilmott, P., Howison, S. and Dewynne, J. (1995): *The Mathematics of Financial Derivatives*, Cambridge, Cambridge University Press.

Chapter 5
CTM and Variance, Volatility, and Covariance and Correlation Swaps for the Classical Heston Model

"Criticism is easy; achievement is difficult". —Winston Churchill.

Abstract In this chapter, we apply the CTM to price variance and volatility swaps for financial markets with underlying assets and variance that follow the classical Heston (Review of Financial Studies 6, 327–343, 1993) model. We also find covariance and correlation swaps for the model. As an application, we provide a numerical example using S&P60 Canada Index to price swap on the volatility (see Swishchuk (2004)).

5.1 Introduction

In the early 1970s, Black and Scholes (1973) made a major breakthrough by deriving pricing formulas for vanilla options written on the stock. The Black-Scholes model assumes that the volatility term is a constant. This assumption is not always satisfied by real-life options as the probability distribution of an equity has a fatter left tail and thinner right tail than the lognormal distribution (see Hull 2000), and the assumption of constant volatility σ in a financial model (such as the original Black-Scholes model) is incompatible with derivatives prices observed in the market.

The issues above have been addressed and studied in several ways, such as:

(i) Volatility is assumed to be a deterministic function at the time: $\sigma \equiv \sigma(t)$ (see Wilmott et al. 1995); Merton (1973) extended the term structure of volatility to $\sigma := \sigma_t$ (deterministic function of time), with the implied volatility for an option of maturity T given by $\hat{\sigma}_T^2 = \frac{1}{T} \int_0^T \sigma_u^2 du$;

(ii) Volatility is assumed to be a function of the time and the current level of the stock price $S(t)$: $\sigma \equiv \sigma(t, S(t))$ (see Hull 2000); the dynamics of the stock price satisfies the following stochastic differential equation:

© The Author 2016
A. Swishchuk, *Change of Time Methods in Quantitative Finance*,
SpringerBriefs in Mathematics, DOI 10.1007/978-3-319-32408-1_5

$$dS(t) = \mu S(t)dt + \sigma(t, S(t))S(t)dW_1(t),$$

where $W_1(t)$ is a standard Wiener process;

(iii) The time variation of the volatility involves an additional source of random-ness, besides $W_1(t)$, represented by $W_2(t)$, and is given by

$$d\sigma(t) = a(t, \sigma(t))dt + b(t, \sigma(t))dW_2(t),$$

where $W_2(t)$ and $W_1(t)$ (the initial Wiener process that governs the price process) may be correlated (see Buff 2002; Hull and White 1987; Heston 1993);

(iv) The volatility depends on a random parameter x such as $\sigma(t) \equiv \sigma(x(t))$, where $x(t)$ is some random process (see Griego and Swishchuk 2000; Swishchuk 1995, 2000; Swishchuk and Kalemanova 2000);

(v) Another approach is connected with stochastic volatility, namely, an uncertain volatility scenario (see Buff 2002). This approach is based on the uncertain volatility model developed in Avellaneda et al. (1995), where a concrete volatility surface is selected among a candidate set of volatility surfaces. This approach addresses the sensitivity question by computing an upper boundary for the value of the portfolio under arbitrary candidate volatility, and this is achieved by choosing the local volatility $\sigma(t, S(t))$ among two extreme values σ_{min} and σ_{max} such that the value of the portfolio is maximized locally;

(vi) The volatility $\sigma(t, S_t)$ depends on $S_t := S(t + \theta)$ for $\theta \in [-\tau, 0]$, namely, stochastic volatility with delay (see Kazmerchuk et al. 2005);

In approach (i), the volatility coefficient is independent of the current level of the underlying stochastic process $S(t)$. This is a deterministic volatility model, and the special case where σ is a constant reduces to the well-known Black-Scholes model that suggests changes in stock prices are lognormal distributed. But the empirical test by Bollerslev (1986) seems to indicate otherwise. One explanation for this problem of a lognormal model is the possibility that the variance of $\log(S(t)/S(t-1))$ changes randomly. This motivated the work of Chesney and Scott (1989), where the prices are analyzed for European options using the modified Black-Scholes model of foreign currency options and a random variance model. In their works the results of Hull and White (1987), Scott (1987), and Wiggins (1987) were used in order to incorporate randomly changing variance rates.

In the approach (ii), several ways have been developed to derive the corresponding Black-Scholes formula: one can obtain the formula by using stochastic calculus and, in particular, Itô's formula (e.g. see Øksendal 1998).

A generalized volatility coefficient of the form $\sigma(t, S(t))$ is said to be *level dependent*. Because volatility and asset price are perfectly correlated, we have only one source of randomness given by $W_1(t)$. A time- and level-dependent volatility coefficient makes the arithmetic more challenging and usually precludes the existence of a closed-form solution. However, the *arbitrage argument* based on portfolio replication and a completeness of the market remains unchanged.

The situation becomes different if the volatility is influenced by a second "non-tradable" source of randomness. This is addressed in approaches (iii), (iv), and (v). We usually obtain in these approaches a *stochastic volatility model*, which is general enough to include the deterministic model as a special case. The concept of stochastic volatility was introduced by Hull and White (1987), and subsequent development includes the work of Wiggins (1987), Johnson and Shanno (1987), Scott (1987), Stein and Stein (1991), and Heston (1993). We also refer to Frey (1997) for an excellent survey on level-dependent and stochastic volatility models. We should mention that the approach (iv) is taken by, for example, Griego and Swishchuk (2000).

Hobson and Rogers (1998) suggested a new class of nonconstant volatility models, which can be extended to include the aforementioned level-dependent model and share many characteristics with the stochastic volatility model. The volatility is nonconstant and can be regarded as an endogenous factor in the sense that it is defined in terms of the *past behaviour of the stock price*. This is done in such a way that the price and volatility form a multidimensional Markov process. Volatility swaps are forward contracts on future realized stock volatility, and variance swaps are similar contracts on variance, the square of the future volatility. Both these instruments provide an easy way for investors to gain exposure to the future level of volatility.

In this chapter we use the CTM to the study stochastic volatility model, the Heston (1993) model, to price variance and volatility swaps. The Heston asset process has a variance σ_t^2 that follows a Cox et al. (1985) process. We find some analytically close forms for expectation and variance of the realized, both continuous and discrete sampled, variance, which are needed for the study of variance and volatility swaps and the price of pseudo-variance, pseudo-volatility. These problems were proposed by He and Wang (2002) for financial markets with deterministic volatility as a function of time. This approach was first applied to the study of stochastic stability of Cox-Ingersoll-Ross process in Swishchuk and Kalemanova (2000).

The same expressions for $E[V]$ and for $Var[V]$ (like in the present chapter) were obtained by Brockhaus and Long (2000) using another analytical approach. Most articles on volatility products focus on the relatively straightforward variance swaps. They take the subject further with a simple model of volatility swaps.

We also study covariance and correlation swaps for security markets with two underlying assets with stochastic volatilities.

As an application of our analytical solutions, we provide a numerical example using *S&P60* Canada Index to price swap on the volatility.

5.2 Variance and Volatility Swaps

Volatility swaps are forward contracts on future realized stock volatility; variance swaps are similar contracts on variance, the square of the future volatility; both these instruments provide an easy way for investors to gain exposure to the future level of volatility.

A stock's volatility is the simplest measure of its riskiness or uncertainty. Formally, the volatility $\sigma_R(S)$ is the annualized standard deviation of the stock's returns during the period of interest, where the subscript R denotes the observed or "realized" volatility for the stock S. The easy way to trade volatility is to use volatility swaps, sometimes called realized volatility forward contracts, because they provide pure exposure to volatility (and only to volatility) (see Demeterfi et al. 1999).

A stock *volatility swap* is a forward contract on the annualized volatility. Its payoff, at expiration, is equal to

$$N(\sigma_R(S) - K_{vol}),$$

where $\sigma_R(S)$ is the realized stock volatility (quoted in annual terms) over the life of the contract,

$$\sigma_R(S) := \sqrt{\frac{1}{T} \int_0^T \sigma_s^2 ds},$$

where σ_t is a stochastic stock volatility, K_{vol} is the annualized volatility delivery price, and N is the notional amount of swap in dollars per annualized volatility point. The holder of a volatility swap at expiration receives N dollars for every point by which the stock's realized volatility σ_R has exceeded the volatility delivery price K_{vol}. The holder is swapping a fixed volatility K_{vol} for the actual (floating) future volatility σ_R. We note that usually $N = \alpha I$, where α is a converting parameter such as 1 per volatility square, and I is a long-short index (+1 for long and -1 for short).

Although market participants speak of volatility, it is variance, or volatility squared, that has more fundamental significance (see Demeterfi et al. 1999).

A *variance swap* is a forward contract on annualized variance, the square of the realized volatility. Its payoff at expiration is equal to

$$N(\sigma_R^2(S) - K_{var}),$$

where $\sigma_R^2(S)$ is the realized stock variance(quoted in annual terms) over the life of the contract,

$$\sigma_R^2(S) := \frac{1}{T} \int_0^T \sigma_s^2 ds,$$

where K_{var} is the delivery price for variance, and N is the notional amount of the swap in dollars per annualized volatility point squared. The holder of variance swap at expiration receives N dollars for every point by which the stock's realized variance $\sigma_R^2(S)$ has exceeded the variance delivery price K_{var}.

Therefore, pricing the variance swap reduces to calculating the realized volatility square.

Valuing a variance forward contract or swap is no different from valuing any other derivative security. The value of a forward contract P on future realized variance with strike price K_{var} is the expected present value of the future payoff in the risk-neutral world:

$$P = E\{e^{-rT}(\sigma_R^2(S) - K_{var})\},$$

where r is the risk-free discount rate corresponding to the expiration date T, and E denotes the expectation.

Thus, for calculating variance swaps, we need to know only $E\{\sigma_R^2(S)\}$, namely, the mean value of the underlying variance.

To calculate volatility swaps, we need more. From the Brockhaus and Long (2000) approximation (which is used in the second-order Taylor expansion for function $\sqrt{x}$), we have (see also Javaheri et al. 2002, p.16)

$$E\{\sqrt{\sigma_R^2(S)}\} \approx \sqrt{E\{V\}} - \frac{Var\{V\}}{8E\{V\}^{3/2}},$$

where $V := \sigma_R^2(S)$ and $\frac{Var\{V\}}{8E\{V\}^{3/2}}$ is the convexity adjustment.

Thus, to calculate volatility swaps, we need both $E\{V\}$ and $Var\{V\}$.

The realized continuously sampled variance is defined in the following way:

$$V := Var(S) := \frac{1}{T} \int_0^T \sigma_t^2 dt.$$

The realized discrete sampled variance is defined as

$$Var_n(S) := \frac{n}{(n-1)T} \sum_{i=1}^{n} \log^2 \frac{S_{t_i}}{S_{t_{i-1}}},$$

where we neglected the following term $\frac{1}{n} \sum_{i=1}^{n} \log \frac{S_{t_i}}{S_{t_{i-1}}}$, since we assume that the mean of the returns is of the order $\frac{1}{n}$ and can be neglected. The scaling by $\frac{n}{T}$ ensures that these quantities annualized (daily) if the maturity T is expressed in years (days).

$Var_n(S)$ is an unbiased variance estimation for σ_t. It can be shown that (see Brockhaus and Long 2000)

$$V := Var(S) = \lim_{n \to +\infty} Var_n(S).$$

Realized discrete sampled volatility is given by

$$Vol_n(S) := \sqrt{Var_n(S)}.$$

Realized continuously sampled volatility is defined as

$$Vol(S) := \sqrt{Var(S)} = \sqrt{V}.$$

The expressions for V, $Var_n(S)$, and $Vol(S)$ are used for the calculation of variance and volatility swaps.

5.3 Variance and Volatility Swaps for the Heston Model of Security Markets

5.3.1 The Stochastic Volatility Model

Let $(\Omega, \mathscr{F}, \mathscr{F}_t, P)$ be a probability space with the filtration $\mathscr{F}_t$, $t \in [0, T]$.

Assume that the underlying asset S_t in the risk-neutral world and variance follow the following Heston (1993) model:

$$\begin{cases} \frac{dS_t}{S_t} = r_t dt + \sigma_t dw_t^1 \\ d\sigma_t^2 = k(\theta^2 - \sigma_t^2)dt + \gamma \sigma_t dw_t^2, \end{cases} \tag{5.1}$$

where r_t is the deterministic interest rate, σ_0 and θ are short and long volatilities, $k > 0$ is a reversion speed, $\gamma > 0$ is a volatility (of volatility) parameter, and w_t^1 and w_t^2 are independent standard Wiener processes.

The Heston asset process has a variance σ_t^2 that follows the Cox et al. (1985) process, described by the second equation in (5.1).

If the volatility σ_t follows the Ornstein-Uhlenbeck process (e.g. see Øksendal 1998), then Ito's lemma shows that the variance σ_t^2 follows the process described exactly by the second equation in (5.1).

5.3.2 Explicit Expression for σ_t^2

In this section we propose a new probabilistic approach to solve the equation for the variance σ_t^2 in (5.1) explicitly, using a change of time method (see Ikeda and Watanabe 1981).

Define the following process:

$$v_t := e^{kt}(\sigma_t^2 - \theta^2). \tag{5.2}$$

Then, using the Itô formula (see Øksendal (1998)), we obtain the equation for v_t:

$$dv_t = \gamma e^{kt} \sqrt{e^{-kt} v_t + \theta^2} dw_t^2. \tag{5.3}$$

Using a change of time approach to the general equation (see Ikeda and Watanabe 1981)

$$dX_t = \alpha(t, X_t) dw_t^2,$$

we obtain the following solution to the equation (5.3):

$$v_t = \sigma_0^2 - \theta^2 + \tilde{w}^2(\hat{T}_t),$$

or (see (5.2)),

$$\sigma_t^2 = e^{-kt}(\sigma_0^2 - \theta^2 + \tilde{w}^2(\hat{T}_t)) + \theta^2, \tag{5.4}$$

where $\tilde{w}^2(t)$ is an $\mathcal{F}_t$-measurable one-dimensional Wiener process, and $\hat{T}_t$ is an inverse function to T_t such that

$$T_t = \gamma^{-2} \int_0^t \{e^{kT_s}(\sigma_0^2 - \theta^2 + \tilde{w}^2(t)) + \theta^2 e^{2kT_s}\}^{-1} ds.$$

5.3.3 Properties of the Processes $\tilde{w}^2(\hat{T}_t)$ and σ_t^2

The properties of $\tilde{w}^2(\hat{T}_t) := b(t)$ are the following:

$$Eb(t) = 0; \tag{5.5}$$

$$E(b(t))^2 = \gamma^2 \{\frac{e^{kt} - 1}{k}(\sigma_0^2 - \theta^2) + \frac{e^{2kt} - 1}{2k}\theta^2\}; \tag{5.6}$$

$$Eb(t)b(s) = \gamma^2 \{\frac{e^{k(t \wedge s)} - 1}{k}(\sigma_0^2 - \theta^2) + \frac{e^{2k(t \wedge s)} - 1}{2k}\theta^2\}, \tag{5.7}$$

where $t \wedge s := \min(t, s)$.

Using representation (5.4) and properties (5.5)–(5.7) of $b(t)$, we obtain the properties of σ_t^2. Straightforward calculations give us the following results:

$$E\sigma_t^2 = e^{-kt}(\sigma_0^2 - \theta^2) + \theta^2;$$

$$E\sigma_t^2 \sigma_s^2 = \gamma^2 e^{-k(t+s)}\{\frac{e^{k(t \wedge s)} - 1}{k}(\sigma_0^2 - \theta^2)$$

$$+ \frac{e^{2k(t \wedge s)} - 1}{2k}\theta^2\} + e^{-k(t+s)}(\sigma_0^2 - \theta^2)^2 \tag{5.8}$$

$$+ e^{-kt}(\sigma_0^2 - \theta^2)\theta^2 + e^{-ks}(\sigma_0^2 - \theta^2)\theta^2 + \theta^4.$$

5.3.4 Valuing Variance and Volatility Swaps

From formula (5.8) we obtain the mean value for V:

$$E\{V\} = \frac{1}{T} \int_0^T E\sigma_t^2 dt$$

$$= \frac{1}{T} \int_0^T \{e^{-kt}(\sigma_0^2 - \theta^2) + \theta^2\} dt \tag{5.9}$$

$$= \frac{1 - e^{-kT}}{kT}(\sigma_0^2 - \theta^2) + \theta^2.$$

The same expression for $E[V]$ may be found in Brockhaus and Long (2000). Substituting $E[V]$ from (5.9) into formula

$$P = e^{-rT}(E\{\sigma_R^2(S)\} - K_{var}) \tag{5.10}$$

we obtain the value of the variance swap.

Variance for V equals

$$Var(V) = EV^2 - (EV)^2.$$

From (5.9) we have

$$(EV)^2 = \frac{1 - 2e^{-kT} + e^{-2kT}}{k^2 T^2}(\sigma_0^2 - \theta^2)^2 + \frac{2(1 - e^{-kT})}{kT}(\sigma_0^2 - \theta^2)\theta^2 + \theta^4. \quad (5.11)$$

Second moment may found by using formula (5.8):

$$\begin{aligned}
EV^2 &= \frac{1}{T^2} \int_0^T \int_0^T E\sigma_t^2 \sigma_s^2 \, dt ds \\
&= \frac{\gamma^2}{T^2} \int_0^T \int_0^T e^{-k(t+s)} \{ \frac{e^{k(t \wedge s)} - 1}{k}(\sigma_0^2 - \theta^2) + \frac{e^{2k(t \wedge s)} - 1}{2k} \theta^2 \} dt ds \quad (5.12) \\
&\quad + \frac{1 - 2e^{-kT} + e^{-2kT}}{k^2 T^2}(\sigma_0^2 - \theta^2)^2 + \frac{2(1 - e^{-kT})}{kT}(\sigma_0^2 - \theta^2)\theta^2 + \theta^4.
\end{aligned}$$

Taking into account (5.11) and (5.12), we obtain

$$\begin{aligned}
Var(V) &= EV^2 - (EV)^2 \\
&= \frac{\gamma^2}{T^2} \int_0^T \int_0^T e^{-k(t+s)} \{ \frac{e^{k(t \wedge s)} - 1}{k}(\sigma_0^2 - \theta^2) + \frac{e^{2k(t \wedge s)} - 1}{2k} \theta^2 \} dt ds.
\end{aligned}$$

After calculations are completed for the last expression, we obtain the following expression for the variance of V :

$$\begin{aligned}
Var(V) = \frac{\gamma^2 e^{-2kT}}{2k^3 T^2} &[(2e^{2kT} - 4e^{kT} kT - 2)(\sigma_0^2 - \theta^2) \\
&+ (2e^{2kT} kT - 3e^{2kT} + 4e^{kT} - 1)\theta^2].
\end{aligned} \quad (5.13)$$

A similar expression for $Var[V]$ may be found in Brockhaus and Long (2000). Substituting EV from (5.9) and $Var(V)$ from (5.13) into the formulav

$$P = \{ e^{-rT}(E\{\sigma_R(S)\} - K_{var}) \} \quad (5.14)$$

with

$$E\{\sigma_R(S)\} = E\{\sqrt{\sigma_R^2(S)}\} \approx \sqrt{E\{V\}} - \frac{Var\{V\}}{8E\{V\}^{3/2}},$$

we obtain the value of volatility swap.

5.3.5 The Calculation of $E\{V\}$ in a Discrete Case

The realized discrete sampled variance is

$$Var_n(S) := \frac{n}{(n-1)T} \sum_{i=1}^n \log^2 \frac{S_{t_i}}{S_{t_{i-1}}},$$

where we neglected the following term $\frac{1}{n}\sum_{i=1}^{n}\log\frac{S_{t_i}}{S_{t_{i-1}}}$ for simplicity reason only. We note that

$$\log\frac{S_{t_i}}{S_{t_{i-1}}} = \int_{t_{i-1}}^{t_i}(r_t - \sigma_t^2/2)dt + \int_{t_{i-1}}^{t_i}\sigma_t dw_t^1.$$

$$E\{Var_n(S)\} = \frac{n}{(n-1)T}\sum_{i=1}^{n}E\{\log^2\frac{S_{t_i}}{S_{t_{i-1}}}\}.$$

$$E\{\log^2\frac{S_{t_i}}{S_{t_{i-1}}}\} = (\int_{t_{i-1}}^{t_i}r_t dt)^2 - \int_{t_{i-1}}^{t_i}r_t dt\int_{t_{i-1}}^{t_i}E\sigma_t^2 dt$$

$$+\frac{1}{4}\int_{t_{i-1}}^{t_i}\int_{t_{i-1}}^{t_i}E\sigma_t^2\sigma_s^2 dtds$$

$$-E(\int_{t_{i-1}}^{t_i}\sigma_t^2 dt\int_{t_{i-1}}^{t_i}\sigma_t dw_t^1) + \int_{t_{i-1}}^{t_i}E\sigma_t^2 dt.$$

We know the expressions for $E\sigma_t^2$ and for $E\sigma_t^2\sigma_s^2$, and the fourth expression is equal to zero. Hence, we can easily calculate all the expressions above and, thus, $E\{Var_n(S)\}$ and variance swap in this case.

Remark 1. Some expressions for the price of the realized discrete sampled variance $Var_n(S) := \frac{n}{(n-1)T}\sum_{i=1}^{n}\log^2\frac{S_{t_i}}{S_{t_{i-1}}}$, (or pseudo-variance) were obtained in the Proceedings of the Sixth Annual PIMS Industrial Problems Solving Workshop, PIMS IPSW 6, UBC, Vancouver, Canada, May 27-31, 2002. Editor: J. Macki, University of Alberta, Canada, June, 2002, pp. 45–55.

5.4 Covariance and Correlation Swaps for Two Assets with Stochastic Volatilities

5.4.1 Definitions of Covariance and Correlation Swaps

Options dependent on exchange rate movements, such as those paying in a currency different from the underlying currency, have an exposure to movements of the correlation between the asset and the exchange rate; this risk may be eliminated by using covariance swap.

A *covariance swap* is a covariance forward contact of the underlying rates S^1 and S^2, which payoff at expiration equal to

$$N(Cov_R(S^1, S^2) - K_{cov}),$$

where K_{cov} is a strike price, N is the notional amount, and $Cov_R(S^1, S^2)$ is a covariance between two assets S^1 and S^2.

Logically, a *correlation swap* is a correlation forward contract of two underlying rates S^1 and S^2 which payoff at expiration equal to

$$N(Corr_R(S^1, S^2) - K_{corr}),$$

where $Corr(S^1, S^2)$ is a realized correlation of two underlying assets S^1 and S^2, K_{corr} is a strike price, and N is the notional amount.

Pricing covariance swap, from a theoretical point of view, is similar to pricing variance swaps, since

$$Cov_R(S^1, S^2) = 1/4\{\sigma_R^2(S^1 S^2) - \sigma_R^2(S^1/S^2)\}$$

where S^1 and S^2 are given two assets, $\sigma_R^2(S)$ is a variance swap for underlying assets, and $Cov_R(S^1, S^2)$ is a realized covariance of the two underlying assets S^1 and S^2.

Thus, we need to know variances for $S^1 S^2$ and for S^1/S^2 (see Section 4.2 for details). Correlation $Corr_R(S^1, S^2)$ is defined as

$$Corr_R(S^1, S^2) = \frac{Cov_R(S^1, S^2)}{\sqrt{\sigma_R^2(S^1)}\sqrt{\sigma_R^2(S^2)}},$$

where $Cov_R(S^1, S^2)$ is defined above and $\sigma_R^2(S^1)$ is defined in Section 3.4.

Given two assets S_t^1 and S_t^2 with $t \in [0, T]$, sampled on days $t_0 = 0 < t_1 < t_2 < \dots < t_n = T$ between today and maturity T, the log return for each asset is $R_i^j :=$

$\log(\frac{S_{t_i}^j}{S_{t_{i-1}}^j})$, $\quad i = 1, 2, \dots, n, \quad j = 1, 2.$

Covariance and correlation can be approximated by

$$Cov_n(S^1, S^2) = \frac{n}{(n-1)T} \sum_{i=1}^{n} R_i^1 R_i^2$$

and

$$Corr_n(S^1, S^2) = \frac{Cov_n(S^1, S^2)}{\sqrt{Var_n(S^1)}\sqrt{Var_n(S^2)}},$$

respectively.

5.4.2 Valuing of Covariance and Correlation Swaps

To value covariance swap, we need to calculate the following:

$$P = e^{-rT}(ECov(S^1, S^2) - K_{cov}). \tag{5.15}$$

To calculate $ECov(S^1, S^2)$ we need to calculate $E\{\sigma_R^2(S^1 S^2) - \sigma_R^2(S^1/S^2)\}$ for the assets S^1 and S^2.

Let S_t^i, $i = 1, 2$, be two strictly positive Itô's processes given by the following model:

$$\begin{cases} \frac{dS_t^i}{S_t^i} = \mu_t^i dt + \sigma_t^i dw_t^i, \\ d(\sigma^i)_t^2 = k^i(\theta_i^2 - (\sigma^i)_t^2)dt + \gamma^i \sigma_t^i dw_t^j, \quad i = 1, 2, \quad j = 3, 4, \end{cases} \quad (5.16)$$

where μ_t^i, $i = 1, 2$, are deterministic functions; k^i, θ^i, γ^i, $i = 1, 2$, are defined in similar way as in (5.1), standard Wiener processes w_t^j, $j = 3, 4$, are independent, $[w_t^1, w_t^2] = \rho_t dt$, ρ_t is a deterministic function of time, $[\cdot, \cdot]$ means the quadratic covariance, and standard Wiener processes w_t^i, $i = 1, 2$, and w_t^j, $j = 3, 4$, are independent.

We note that

$$d \ln S_t^i = m_t^i dt + \sigma_t^i dw_t^i, \quad (5.17)$$

where

$$m_t^i := (\mu_t^i - \frac{(\sigma_t^i)^2}{2}), \quad (5.18)$$

and

$$Cov_R(S_T^1, S_T^2) = \frac{1}{T}[\ln S_T^1, \ln S_T^2] = \frac{1}{T}[\int_0^T \sigma_t^1 dw_t^1, \int_0^T \sigma_t^2 dw_t^2] = \frac{1}{T}\int_0^T \rho_t \sigma_t^1 \sigma_t^2 dt. \quad (5.19)$$

Let us show that

$$[\ln S_T^1, \ln S_T^2] = \frac{1}{4}([\ln(S_T^1 S_T^2)] - [\ln(S_T^1/S_T^2)]). \quad (5.20)$$

First, note that

$$d\ln(S_t^1 S_t^2) = (m_t^1 + m_t^2)dt + \sigma_t^+ dw_t^+, \quad (5.21)$$

and

$$d\ln(S_t^1/S_t^2) = (m_t^1 - m_t^2)dt + \sigma_t^- dw_t^-, \quad (5.22)$$

where

$$(\sigma_t^\pm)^2 := (\sigma_t^1)^2 \pm 2\rho_t \sigma_t^1 \sigma_t^2 + (\sigma_t^2)^2, \quad (5.23)$$

and

$$dw_t^\pm := \frac{1}{\sigma_t^\pm}(\sigma_t^1 dw_t^1 \pm \sigma_t^2 dw_t^2). \quad (5.24)$$

Processes $w_t^\pm$ in (5.24) are standard Wiener processes by the Levi-Kunita-Watanabe theorem and $\sigma_t^\pm$ are defined in (5.23).

In this way, from (5.21) and (5.22), we obtain that

$$[\ln(S_t^1 S_t^2)] = \int_0^t (\sigma_s^+)^2 ds = \int_0^t ((\sigma_s^1)^2 + 2\rho_t \sigma_s^1 \sigma_s^2 + (\sigma_s^2)^2)ds, \quad (5.25)$$

and

$$[\ln(S_t^1/S_t^2)] = \int_0^t (\sigma_s^-)^2 ds = \int_0^t ((\sigma_s^1)^2 - 2\rho_t \sigma_s^1 \sigma_s^2 + (\sigma_s^2)^2)ds. \quad (5.26)$$

From (5.20), (5.25), and (5.26), we have the direct of (5.20)

$$[\ln S_T^1, \ln S_T^2] = \frac{1}{4}([\ln(S_T^1 S_T^2)] - [\ln(S_T^1/S_T^2)]). \tag{5.27}$$

Thus, from (5.27) we obtain that (see (5.20) and Section 4.1))

$$Cov_R(S^1, S^2) = 1/4(\sigma_R^2(S^1 S^2) - \sigma_R^2(S^1/S^2)).$$

Returning to the valuation of the covariance swap, we have

$$P = E\{e^{-rT}(Cov(S^1, S^2) - K_{cov}\} = \frac{1}{4}e^{-rT}(E\sigma_R^2(S^1 S^2) - E\sigma_R^2(S^1/S^2) - 4K_{cov}).$$

The problem now has reduced to the same problem as in the Section 5.3, but instead of σ_t^2, we need to take $(\sigma_t^+)^2$ for $S^1 S^2$ and $(\sigma_t^-)^2$ for S^1/S^2 (see (5.23)) and proceed with similar calculations as in Section 5.3.

Remark 2. The results of the Sections 5.2–5.4 were first presented on the Sixth Annual Financial Econometrics Conference "Estimation of Diffusion Processes in Finance", Friday, March 19, 2004, Centre for Advanced Studies in Finance, University of Waterloo, Waterloo, Canada (*Abstract on-line:* http://arts.uwaterloo.ca/ACCT /finance/fec6.htm).

5.5 A Numerical Example: $S\&P60$ Canada Index

In this section, we apply the analytical solutions from Section 5.3 to price a swap on the volatility of the S&P60 Canada Index for five years (January 1997–February 2002).

This data was kindly presented to the author by Raymond Théoret (Université du Québec à Montréal, Montréal, Québec, Canada) and Pierre Rostan (analyst at the R&D Department of Bourse de Montréal and Université du Québec à Montréal, Montréal, Québec, Canada). They calibrated the GARCH parameters from five years of daily historic *S&P60* Canada Index (from January 1997 to February 2002) [see working paper "Pricing volatility swaps: Empirical testing with Canadian data" by Theoret et al. (2002)].

At the end of February 2002, we wanted to price the fixed leg of a volatility swap based on the volatility of the S&P60 Canada Index. The statistics on log returns *S&P60* Canada Index for 5 years (January 1997–February 2002) are presented in Table 5.1.

From the histogram of the S&P60 Canada Index log returns on a 5-year historical period (1,300 observations from January 1997 to February 2002), leptokurtosis may be seen in the histogram. If we take a look at the graph of the S&P60 Canada Index log returns on a 5-year historical period, we may see volatility clustering in the returns series. These facts give us information about the conditional heteroscedasticity. A GARCH(1,1) regression is applied to the series and the result is obtained (see Table 5.2).

Table 5.1 Statistics on log returns *S&P*60 Canada Index

Statistics on log returns *S&P*60 **Canada Index**	
Series:	Log returns *S&P*60 Canada Index
Sample:	1 1300
Observations:	1300
Mean	0.0002
Median	0.0006
Maximum	0.0520
Minimum	−0.1011
Std. dev.	0.0136
Skewness	−0.6657
Kurtosis	7.7873

Table 5.2 Estimation of the GARCH(1,1) process

Estimation of the GARCH(1,1) process				
Dependent variable: log returns of S&P60 Canada Index prices				
Method: ML-ARCH				
Included observations: 1300				
Convergence achieved after 28 observations				
-	**Coefficient:**	**Std. error:**	*z*-**statistic:**	**Prob.**
C	0.000617	0.000338	1.824378	0.0681
Variance equation				
C	2.58E-06	3.91E-07	6.597337	0
ARCH(1)	0.060445	0.007336	8.238968	0
GARCH(1)	0.927264	0.006554	141.4812	0
R-squared	-0.000791	Mean dependent var	-	0.000235
Adjusted R-squared	-0.003108	S.D. dependent var	-	0.013567
S.E. of regression	0.013588	Akaike info criterion	-	-5.928474
Sum squared resid	0.239283	Schwartz criterion	-	-5.912566
Log likelihood	3857.508	Durbin-Watson stat	-	1.886028

This Table 5.2 allows us to generate different input variables to the volatility swap model.

We use the following relationships:

$$\theta = \frac{V}{dt},$$

$$k = \frac{1 - \alpha - \beta}{dt},$$

$$\gamma = \alpha \sqrt{\frac{\xi - 1}{dt}},$$

to calculate the following discrete GARCH(1,1) parameters:

ARCH(1,1) coefficient $\alpha = 0.060445$;
GARCH(1,1) coefficient $\beta = 0.927264$;
the Pearson kurtosis (fourth moment of the drift-adjusted stock return) $\xi = 7.787327$;
long volatility $\theta = 0.05289724$;
$k = 3.09733$;
$\gamma = 2.499827486$;
a short volatility σ_0 equals to 0.01;
Parameter V may be found from the expression $V = \frac{C}{1-\alpha-\beta}$, where $C = 2.58 \times 10^{-6}$ is defined in Table 5.2. Thus, $V = 0.00020991$;
$dt = 1/252 = 0.003968254$.
Now, applying the analytical solutions (5.9) and (5.13) for a swap maturity T of 0.91 year, we find the following values:

$$E\{V\} = \frac{1-e^{-kT}}{kT}(\sigma_0^2 - \theta^2) + \theta^2 = .3364100835,$$

and

$$Var(V) = \frac{\gamma^2 e^{-2kT}}{2k^3 T^2}[(2e^{2kT} - 4e^{kT}kT - 2)(\sigma_0^2 - \theta^2) + (2e^{2kT}kT - 3e^{2kT} + 4e^{kT} - 1)\theta^2] = .0005516049969.$$

The convexity adjustment $\frac{Var\{V\}}{8E\{V\}^{3/2}}$ is equal to .0003533740855.
If the non-adjusted strike is equal to 18.7751%, then the adjusted strike is equal to

$$18.7751\% - 0.03533740855\% = 18.7398\%.$$

This is the fixed leg of the volatility swap for a maturity $T = 0.91$.

Repeating this approach for a series of maturities up to 10 years, we may obtain the plot presented in the Appendix, Figure 5.1 (see S&P60 Canada Index volatility swap).

Figure 5.2 (see Appendix) illustrates the non-adjusted and adjusted volatility for the same series of maturities.

5.6 Appendix: Figures

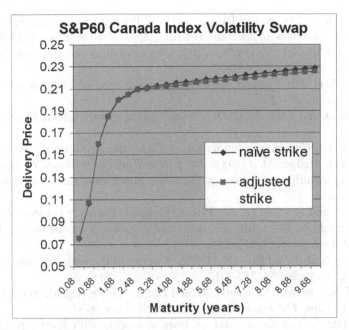

Fig. 5.1 *S&P*60 Canada Index volatility swap

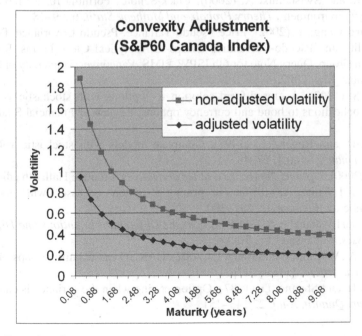

Fig. 5.2 Convexity adjustment

References

Avellaneda, M., Levy, A. and Paras, A. (1995): Pricing and hedging derivative securities in markets with uncertain volatility, *Appl. Math. Finance* 2, 73–88.

Black, F. and Scholes, M. (1973): The pricing of options and corporate liabilities, *J. Political Economy* 81, 637–54.

Bollerslev, T. (1986): Generalized autoregressive conditional heteroscedasticity, *J. Economics* 31, 307–27.

Brockhaus, O. and Long, D. (2000): Volatility swaps made simple, RISK, January, 92–96.

Buff, R. (2002): *Uncertain volatility model. Theory and Applications*. NY: Springer.

Carr, P. and Madan, D. (1998): *Towards a Theory of Volatility Trading*. In the book: Volatility, Risk book publications, http://www.math.nyu.edu/research/carrp/papers/.

Chesney, M. and Scott, L. (1989): Pricing European Currency Options: A comparison of modified Black-Scholes model and a random variance model, *J. Finan. Quantit. Anal.* 24, No3, 267–284.

Cox, J., Ingersoll, J. and Ross, S. (1985): A theory of the term structure of interest rates, Econometrica 53, 385–407.

Demeterfi, K., Derman, E., Kamal, M. and Zou, J. (1999): A guide to volatility and variance swaps, *The Journal of Derivatives*, Summer, 9–32.

Frey, R. (1997): Derivative asset analysis in models with level-dependent and stochastic volatility, *CWI Quarterly* 10, 1–34.

Griego, R. and Swishchuk, A. (2000): Black-Scholes Formula for a Market in a Markov Environment, *Theory Probab. and Mathem. Statit.* 62, 9–18.

He, R. and Wang, Y. (2002): Price Pseudo-Variance, Pseudo Covariance, Pseudo-Volatility, and Pseudo-Correlation Swaps-In Analytical Close Forms, RBC Financial Group, Query Note for 6th ISPW, PIMS, Vancouver, University of British Columbia, May 2002.

Heston, S. (1993): A closed-form solution for options with stochastic volatility with applications to bond and currency options, Review of Financial Studies, 6, 327–343

Hobson, D. and Rogers, L. (1998): Complete models with stochastic volatility, *Math. Finance* 8, no.1, 27–48.

Hull, J. (2000): *Options, futures and other derivatives*, Prentice Hall, 4th edition.

Hull, J., and White, A. (1987): The pricing of options on assets with stochastic volatilities, *J. Finance* 42, 281–300.

Ikeda, N. and Watanabe, S. (1981): *Stochastic Differential Equations and Diffusion Processes*, Kodansha Ltd., Tokyo.

Javaheri, A., Wilmott, P. and Haug, E. (2002): GARCH and volatility swaps, *Wilmott Technical Article*, January, 17p.

Johnson, H. and Shanno, D. (1987): Option pricing when the variance is changing, *J. Finan. Quantit. Anal.* 22,143–151.

Kazmerchuk, Y., Swishchuk, A. and Wu, J. (2005): A continuous-time GARCH model for stochastic volatility with delay, 18p. Canad. Appl. Mathem. Quart., 13, (2), 2005

Merton, R. (1973): Theory of rational option pricing, *Bell Journal of Economic Management Science* 4, 141–183.

Øksendal, B. (1998): *Stochastic Differential Equations: An Introduction with Applications*. NY: Springer.

Scott, L. (1987): Option pricing when the variance changes randomly: theory, estimation and an application, *J. Fin. Quant. Anal.* 22, 419–438.

Stein, E. and Stein, J. (1991): Stock price distributions with stochastic volatility: an analytic approach, *Review Finan. Studies* 4, 727–752.

Swishchuk, A. (1995): Hedging of options under mean-square criterion and with semi-Markov volatility, *Ukrain. Math. J.* 47, No7, 1119–1127.

Swishchuk, A. (2000): *Random Evolutions and Their Applications. New Trends*, Kluwer Academic Publishers, Dordrecht, The Netherlands.

Swishchuk, A. and Kalemanova, A. (2000): Stochastic stability of interest rates with jumps. *Theory probab. & Mathem statist.*, TBiMC Sci. Publ., v.61. (*Preprint online:* www.math.yorku.ca/~aswishch/sample.html)

Swishchuk, A. (2004): Modeling and pricing of variance and volatility swaps for financial markets with stochastic volatilities, *Wilmott Magazine*, Issue 19, September, N.2, 64–73.

Theoret, R., Zabre, L. and Rostan, P. (2002): Pricing volatility swaps: empirical testing with Canadian data. *Working paper*, Centre de Recherche en Gestion, Document 17-2002, July 2002.

Wiggins, J. (1987): Option values under stochastic volatility: Theory and Empirical Estimates, *J. Finan. Econ.* 19, 351–372.

Wilmott, P., Howison, S. and Dewynne, J. (1995): *Option Pricing: Mathematical Models and Computations*. Oxford: Oxford Financial Press.

References

Chapter 6
CTM and the Delayed Heston Model: Pricing and Hedging of Variance and Volatility Swaps

"Better three hours too soon than a minute too late". —William Shakespeare.

Abstract In this chapter, we apply the CTM for pricing and hedging of variance and volatility swaps for the delayed Heston model. We present a variance drift-adjusted version of the Heston model which leads to a significant improvement of the market volatility surface fitting (compared to the classical Heston model). The numerical example we performed with recent market data shows a significant reduction of the average absolute calibration error[1] (calibration on 12 dates ranging from 19 September to 17 October 2011 for the FOREX underlying EURUSD). Our model has two additional parameters compared to the Heston model and can be implemented very easily. It was initially introduced for the purpose of volatility derivative pricing. The main idea behind our model is to take into account some past history of the variance process in its (risk-neutral) diffusion. Using a change of time method for continuous local martingales, we derive a closed formula for the Brockhaus and Long approximation of the volatility swap price in this model. We also consider dynamic hedging of volatility swaps using a portfolio of variance swaps.

6.1 Introduction

The volatility process is an important concept in financial modelling as it quantifies at each time t how likely the modelled asset log return is to vary significantly over some short immediate time period $[t, t + \varepsilon]$. This process can be stochastic or deterministic, e.g. local volatility models in which the (deterministic) volatility depends on time and spot price level. In quantitative finance, we often consider the volatility process $\sqrt{V_t}$ (where V_t is the variance process) to be stochastic, as it allows us to fit

---

[1] The average absolute calibration error is defined to be the average of the absolute values of the differences between market and model implied Black and Scholes volatilities.

© The Author 2016   59
A. Swishchuk, *Change of Time Methods in Quantitative Finance*,
SpringerBriefs in Mathematics, DOI 10.1007/978-3-319-32408-1_6

the observed vanilla option market prices with an acceptable bias as well as to model the risk linked with the future evolution of the volatility smile (which the deterministic model cannot do), namely, the forward smile. Many derivatives are known to be very sensitive to the forward smile. One of the most popular example being the cliquet options (options on future asset performance; e.g. see Kruse and Nögel 2005).

The Heston model (Heston 1993; Heston and Nandi 2000) is one of the most popular stochastic volatility models in the industry as semi-closed formulas for vanilla option prices are available, few (five) parameters need to be calibrated, and it accounts for the mean-reverting feature of the volatility.

One might be willing, in the variance diffusion, to take into account not only its current state but also its past history over some interval $[t - \tau, t]$, where $\tau > 0$ is a constant and is called the delay. Starting from the discrete-time GARCH(1,1) model of Bollerslev (1986), a first attempt in this direction was made in Kazmerchuk et al. (2007), where a non-Markov delayed continuous-time GARCH model was proposed (S_t being the asset price at time t and γ, θ, and α some positive constants). The dynamics considered had the form

$$\frac{dV_t}{dt} = \gamma \theta^2 + \frac{\alpha}{\tau} \ln^2 \left(\frac{S_t}{S_{t-\tau}} \right) - (\alpha + \gamma)V_t. \qquad (6.1)$$

This model was inherited from its discrete-time analogue (where L is a positive integer):

$$\sigma_n^2 = \tilde{\gamma}\tilde{\theta}^2 + \frac{\tilde{\alpha}}{L} \ln^2 \left(\frac{S_{n-1}}{S_{n-1-L}} \right) + (1 - \tilde{\alpha} - \tilde{\gamma})\sigma_{n-1}^2. \qquad (6.2)$$

The parameter θ^2 (resp. γ) can be interpreted as the value of the long-range variance (resp. variance mean-reversion speed) when the delay is equal to 0 (we will see that introducing a delay modifies the value of these two model features). α is a continuous-time equivalent to the variance ARCH(1,1) autoregressive coefficient. In fact, we can interpret the right-hand side of the diffusion equation (6.2) as the sum of two terms:

- The delay-free term $\gamma(\theta^2 - V_t)$, which accounts for the mean-reverting feature of the variance process.
- $\alpha \left(\frac{1}{\tau} \ln^2 \left(\frac{S_t}{S_{t-\tau}} \right) - V_t \right)$ which is a purely (noisy) delay term, i.e. one that vanishes when $\tau \to 0$ and takes into account the past history of the variance (via the term $\ln \left(\frac{S_t}{S_{t-\tau}} \right)$). The autoregressive coefficient α can be seen as the amplitude of this purely delay term.

In Swishchuk (2005) and Swishchuk and Li (2011), the authors point out the importance of incorporating the real-world $P-$drift $d_P(t, \tau) := \int_{t-\tau}^{t} (\mu - \frac{1}{2}V_u)du$ of $\ln \left(\frac{S_t}{S_{t-\tau}} \right)$ in the model, where μ stands for the real-world $\mathbb{P}-$drift of the stock price S_t. The transformed variance dynamics then is

$$\frac{dV_t}{dt} = \gamma\theta^2 + \frac{\alpha}{\tau}\left[\ln\left(\frac{S_t}{S_{t-\tau}}\right) - d_{\mathbb{P}}(t,\tau)\right]^2 - (\alpha+\gamma)V_t. \tag{6.3}$$

The latter diffusion (6.3) was introduced in Swishchuk (2005) and Kazmerchuk et al. (2005), and the proposed model was proved to be complete and to account for the mean-reverting feature of the volatility process. This model is also non-Markov as the past history $(V_u)_{u\in[t-\tau,t]}$ of the variance appears in its diffusion equation via the term $\ln\left(\frac{S_t}{S_{t-\tau}}\right)$, as shown in Swishchuk (2005). Following this approach, a series of papers was published by one of the authors (Swishchuk 2005) focusing on the pricing of variance swaps in this delayed framework: one-factor stochastic volatility with delay has been presented in Swishchuk (2005); multifactor stochastic volatility with delay in Swishchuk (2006), one-factor stochastic volatility with delay and jumps in Swishchuk and Li (2011), and finally local Levy-based stochastic volatility with delay in Swishchuk and Malenfant (2011).

Other papers related to the concept of delay are also of interest. For example, Kind et al. (1991) obtained a diffusion approximation result for processes satisfying some equations with past-dependent coefficients, with application to option pricing. Arriojas et al. (2007) derived a Black and Scholes formula for call options assuming the stock price follows a stochastic delay differential equation (SDDE). Mohammed and Bell have also published a series of papers in which they investigate various properties of SDDE (e.g. see Bell and Mohammed 1991, 1995).

Unfortunately, the model (6.3) doesn't lead to (semi-)closed formulas for the vanilla options, making it difficult to use for practitioners willing to calibrate on vanilla market prices. Nevertheless, one can notice that the Heston model and the delayed continuous-time GARCH model (6.3) are very similar in the sense that the expected values of the variances are the same-when we make the delay tend to 0 in (6.3). As mentioned before, the Heston framework is very convenient, and therefore it is naturally tempting to adjust the Heston dynamics in order to incorporate the delay introduced in (6.3). In this way, we considered at first adjusting the Heston drift by a deterministic function of time so that the expected value of the variance under the delayed Heston model is equal to the one under the delayed GARCH model (6.3). In addition to making our delayed Heston framework coherent with (6.3), this construction makes the variance process diffusion dependent not on its past history $(V_u)_{u\in[t-\tau,t]}$, but on the past history of its risk-neutral expectation $(\mathbb{E}_0^{\mathbb{Q}}(V_u))_{u\in[t-\tau,t]}$, preserving the Markov feature of the Heston model (where we denote $\mathbb{E}_t^{\mathbb{Q}}(\cdot) := \mathbb{E}^{\mathbb{Q}}(\cdot|\mathscr{F}_t)$ for some filtration $(\mathscr{F}_t)_{t\geq 0}$). The purpose of sections 6.2 and 6.3 is to present the delayed Heston model as well as some calibration results on call option prices, with a comparison to the Heston model. In sections 6.4 and 6.5, we will consider the pricing and hedging of volatility and variance swaps in this model see (Swishchuk and Vadori (2014)).

Volatility and variance swaps are contracts whose payoffs depend (respectively, convexly and linearly) on the realized variance of the underlying asset over some specified time interval. They provide pure exposure to volatility and therefore make it a tradable market instrument. Variance swaps are even considered by some practitioners to be vanilla derivatives. The most commonly traded variance swaps are discretely sampled and have a payoff $P_n^V(T)$ at maturity T of the form

$$P_n^V(T) = N \left[\frac{252}{n} \sum_{i=0}^{n} \ln^2 \left(\frac{S_{i+1}}{S_i} \right) - K_{var} \right],$$

where S_i is the asset spot price on the fixing time $t_i \in [0, T]$ (usually there is one fixing time each day, but there could be more or less), N is the notional amount of the contract (in currency per unit of variance), and K_{var} is the strike specified in the contract. The corresponding volatility swap payoff $P_n^v(T)$ is given by

$$P_n^v(T) = N \left[\sqrt{\frac{252}{n} \sum_{i=0}^{n} \ln^2 \left(\frac{S_{i+1}}{S_i} \right)} - K_{vol} \right].$$

One can also consider continuously sampled volatility and variance swaps (on which we will focus in this article), in which payoffs are, respectively, defined as the limit when $n \to +\infty$ of their discretely sampled versions. Formally, if we denote $(V_t)_{t \geq 0}$ the stochastic volatility process of our asset, adapted to some Brownian filtration $(\mathscr{F}_t)_{t \geq 0}$, then the continuously sampled realized variance V_R from the initiation date of the contract $t = 0$ to the maturity date $t = T$ is given by $V_R = \frac{1}{T} \int_0^T V_s ds$. The fair variance strike K_{var} is calculated such that the initial value of the contract is 0 and therefore is given by

$$\mathbb{E}_0^{\mathbb{Q}} \left[e^{-rT} (V_R - K_{var}) \right] = 0 \Rightarrow K_{var} = \mathbb{E}_0^{\mathbb{Q}}(V_R).$$

In the same way, the fair volatility strike K_{vol} is given by

$$\mathbb{E}_0^{\mathbb{Q}} \left[e^{-rT} (\sqrt{V_R} - K_{vol}) \right] = 0 \Rightarrow K_{vol} = \mathbb{E}_0^{\mathbb{Q}}(\sqrt{V_R}).$$

The volatility swap fair strike might be difficult to compute explicitly as we have to compute the expectation of a square root. In Brockhaus and Long (2000), the following approximation-based on a Taylor expansion-was proposed to compute the expected value of the square root of an almost surely non-negative random variable Z:

$$\mathbb{E}(\sqrt{Z}) \approx \sqrt{\mathbb{E}(Z)} - \frac{Var(Z)}{8 \mathbb{E}(Z)^{\frac{3}{2}}}. \tag{6.4}$$

We will refer to this approximation in our paper as the Brockhaus and Long approximation.

There exists a vast literature on volatility and variance swaps. In the following lines, we provide a selection of papers covering important topics. Carr and

Lee (2009) provide an overview of the current market of volatility derivatives. They survey the early literature on the subject. They also provide relatively simple proofs of some fundamental results related to variance swaps and volatility swaps. Pricing of variance swaps for a one-factor stochastic volatility is presented in Swishchuk (2004). Variance and volatility swaps in energy markets are considered in Swishchuk (2013). Broadie and Jain (2008) cover pricing and dynamic hedging of volatility derivatives in the Heston model. Moreover, various papers deal with the VIX-the Chicago Board Options Exchange Market Volatility Index-which is a popular measure of the one-month implied volatility on the S&P 500 index (e.g. see Zhang and Zhu 2006; Hao and Zhang 2013 or Filipovic 2013).

6.2 Presentation of the Delayed Heston Model

Throughout this chapter, we will assume a constant risk-free rate r, a dividend yield q, and a finite time horizon T. We fix $(\Omega, \mathscr{F}, \mathbb{P})$ a probability space and we consider a stock whose price process is denoted by $(S_t)_{t \geq 0}$. We let $\mathbb{Q}$ be a risk-neutral measure and we let $(Z_t^{\mathbb{Q}})_{t \geq 0}$ and $(W_t^{\mathbb{Q}})_{t \geq 0}$ be two correlated standard Brownian motions on $(\Omega, \mathscr{F}, \mathbb{Q})$. We consider the natural filtration associated to these Brownian motions $\mathscr{F}_t := \sigma(Z_t^{\mathbb{Q}}, W_t^{\mathbb{Q}})$ and we denote $\mathbb{E}_t^{\mathbb{Q}}(\cdot) := \mathbb{E}^{\mathbb{Q}}(\cdot | \mathscr{F}_t)$ and $Var_t^{\mathbb{Q}}(\cdot) := Var^{\mathbb{Q}}(\cdot | \mathscr{F}_t)$.

We assume the following risk-neutral $\mathbb{Q}-$ stock price dynamics:

$$dS_t = (r-q)S_t dt + S_t \sqrt{V_t} dZ_t^{\mathbb{Q}}. \tag{6.5}$$

The well-known Heston model has the following $\mathbb{Q}-$dynamics for the variance V_t:

$$dV_t = \gamma(\theta^2 - V_t)dt + \delta\sqrt{V_t}dW_t^{\mathbb{Q}}, \tag{6.6}$$

where θ^2 is the long-range variance, γ is the variance mean-reversion speed, δ is the volatility of the variance, and ρ is the Brownian correlation coefficient ($\langle W^{\mathbb{Q}}, Z^{\mathbb{Q}} \rangle_t = \rho t$). We also assume $S_0 = s_0$ a.e. and $V_0 = v_0$ a.e., for some positive constants v_0 and s_0.

As explained in the introduction, the following delayed continuous-time GARCH dynamics have been introduced for the variance in Swishchuk (2005):

$$\frac{dV_t}{dt} = \gamma\theta^2 + \frac{\alpha}{\tau}\left[\int_{t-\tau}^t \sqrt{V_s}dZ_s^{\mathbb{Q}} - (\mu - r)\tau\right]^2 - (\alpha + \gamma)V_t, \tag{6.7}$$

where μ stands for the real-world $\mathbb{P}-$drift of the stock price S_t. We notice that θ^2 (resp. γ) has been defined in the introduction for the delayed continuous-time GARCH model as the value of the long-range variance (resp. variance mean-reversion speed) when $\tau = 0$; therefore it has the same meaning as the Heston long-range variance (resp. variance mean-reversion speed). That is why we use the same notations in both models.

We can see that the two models are very similar. Indeed, they both give the same expected value for V_t as the delay goes to 0 in (6.7), namely, $\theta^2 + (V_0 - \theta^2)e^{-\gamma t}$. The idea here is to adjust the Heston dynamics (6.6) in order to account for the delay introduced in (6.7). Our approach is to adjust the drift by a deterministic function of time so that the expected value of V_t under the adjusted Heston model is the same as under (6.7). This approach can be seen as a correction by a pure delay term of amplitude α of the Heston drift by a deterministic function in order to account for the delay.

Namely, we assume the adjusted Heston dynamics:

$$dV_t = \left[\gamma(\theta^2 - V_t) + \varepsilon_\tau(t)\right]dt + \delta\sqrt{V_t}dW_t^{\mathbb{Q}}, \tag{6.8}$$

$$\varepsilon_\tau(t) := \alpha\tau(\mu - r)^2 + \frac{\alpha}{\tau}\int_{t-\tau}^t v_s ds - \alpha v_t, \tag{6.9}$$

with $v_t := \mathbb{E}_0^{\mathbb{Q}}(V_t)$. It was shown in Swishchuk (2005) that v_t solves the following equation:

$$\frac{dv_t}{dt} = \gamma\theta^2 + \alpha\tau(\mu - r)^2 + \frac{\alpha}{\tau}\int_{t-\tau}^t v_s ds - (\alpha + \gamma)v_t, \tag{6.10}$$

and that we have the following expression for v_t

$$v_t = \theta_\tau^2 + (V_0 - \theta_\tau^2)e^{-\gamma_\tau t}, \tag{6.11}$$

with

$$\theta_\tau^2 := \theta^2 + \frac{\alpha\tau(\mu - r)^2}{\gamma}. \tag{6.12}$$

By (6.11) and (6.15) (see below), we have $\lim_{t \to \infty} v_t = \theta_\tau^2$, and therefore the parameter θ_τ^2 can be interpreted as the adjusted value of the limit to which v_t tends to as $t \to \infty$, which has been (positively) shifted from its original value θ^2 because of the introduction of delay. We have $\theta_\tau^2 \to \theta^2$ when $\tau \to 0$, which is coherent. We will see below that we can interpret the parameter $\gamma_\tau > 0$ as the adjusted mean-reversion speed. This parameter is given in Swishchuk (2005) by a (nonzero) solution to the following equation:

$$\gamma_\tau = \alpha + \gamma + \frac{\alpha}{\gamma_\tau\tau}(1 - e^{\gamma_\tau\tau}). \tag{6.13}$$

By (6.9), (6.11), and (6.13), we get an explicit expression for the drift adjustment:

$$\varepsilon_\tau(t) = \alpha\tau(\mu - r)^2 + (V_0 - \theta_\tau^2)(\gamma - \gamma_\tau)e^{-\gamma_\tau t}. \tag{6.14}$$

The following simple property gives us some information about the correction term $\varepsilon_\tau(t)$ and the parameter γ_τ, which will be useful for interpretation purpose and in the derivation of the semi-closed formulas for call options in Appendix 6.6. Indeed, given (6.15) and (6.11), the parameter γ_τ can be interpreted as the adjusted variance mean-reversion speed because it quantifies the speed at which v_t tends to θ_τ^2 as $t \to \infty$, and we have by using a Taylor expansion in (6.13) that $\gamma_\tau \to \gamma$ when $\tau \to 0$, which is coherent.

Property 1: γ_τ is the unique solution to (6.13) and

$$0 < \gamma_\tau < \gamma, \quad \lim_{\tau \to 0} \sup_{t \in \mathbb{R}^+} |\varepsilon_\tau(t)| = 0. \tag{6.15}$$

Proof: Let's show $\gamma_\tau \geq 0$. If $\gamma_\tau, < 0$ then by (6.13), we have $\alpha + \gamma + \frac{\alpha}{\gamma_\tau \tau}(1 - e^{\gamma_\tau \tau}) < 0$, i.e. $1 - e^{\gamma_\tau \tau} + \gamma_\tau \tau > -\frac{\gamma}{\alpha}\gamma_\tau \tau$. But $\tau > 0$ so $\exists x_0 > 0$ s.t. $1 - e^{-x_0} - x_0 > \frac{\gamma}{\alpha}x_0$. A simple study shows that is impossible whenever $\frac{\gamma}{\alpha} \geq 0$, which is what we have by assumption. Therefore $\gamma_\tau \geq 0$, and in fact $\gamma_\tau > 0$ since it is a nonzero solution of (6.13). If $\gamma \leq \gamma_\tau$ then by (6.13) $\gamma_\tau \tau + 1 - e^{\gamma_\tau \tau} \geq 0$. But $\gamma_\tau \tau > 0$; therefore $\exists x_0 > 0$ s.t. $x_0 + 1 - e^{x_0} \geq 0$. A simple study shows that it is impossible. The uniqueness comes from a similar simple study. Now, because $\gamma_\tau > 0$, we have $\sup_{t \in \mathbb{R}^+} |\varepsilon_\tau(t)| \leq \alpha \tau(\mu - r)^2 + |(V_0 - \theta_\tau^2)(\gamma - \gamma_\tau)|$ and $(V_0 - \theta_\tau^2)(\gamma - \gamma_\tau) = \circ(1)$ by (6.13). So $\lim_{\tau \to 0} \alpha \tau(\mu - r)^2 + |(V_0 - \theta_\tau^2)(\gamma - \gamma_\tau)| = 0$.

Using (6.14) and (6.12), we can rewrite (6.8) as a time-dependent Heston model with time-dependent long-range variance $\tilde{\theta}_t^2$:

$$dV_t = \gamma(\tilde{\theta}_t^2 - V_t)dt + \delta\sqrt{V_t}dW_t^\mathbb{Q}, \tag{6.16}$$

$$\tilde{\theta}_t^2 := \theta_\tau^2 + (V_0 - \theta_\tau^2)\frac{(\gamma - \gamma_\tau)}{\gamma}e^{-\gamma_\tau t}. \tag{6.17}$$

The parameter θ_τ^2 is-as we mentioned above-the adjusted value of the limit towards which v_t tends to as $t \to \infty$. For this reason, it is coherent that it is also the limiting value of the time-dependent long-range variance $\tilde{\theta}_t^2$ as $t \to \infty$ (by (6.17) and (6.15)).

6.3 Calibration on Call Option Prices and Comparison to the Heston Model

Following Kahl and Jäckel (2005) and Mikhailov and Noegel (2005), it is possible to get semi-closed formulas for call options in our delayed Heston model. Indeed, our model is a time-dependent Heston model with time-dependent long-range variance $\tilde{\theta}_t^2$. We refer to Appendix 6.6 for the procedure to derive such semi-closed formulas.

We perform our calibration on 30 September 2011 for underlying EURUSD on the whole volatility surface (maturities from 1M to 10Y, strikes ATM, 25D Call/Put, 10D Call/Put). The implied volatility surface, the zero coupon curves EUR vs. EURIBOR 6M and USD vs. LIBOR 3M, and the spot price are taken from Bloomberg (mid-prices). The drift $\mu = 0.0188$ is estimated from 7.5Y of daily close prices (source: www.forexrate.co.uk).

The calibration procedure is a least-squares minimization procedure that we perform via MATLAB (function *lsqnonlin* that uses a trust-region-reflective algorithm). The Heston integral (6.83) is computed via the MATLAB function *quadl* that uses a recursive adaptive Lobatto quadrature. The integral $\int_0^t e^{-\gamma_\tau s} D(s, u) ds$ in (6.81) is computed via a composite Simpson's rule with 100 points.

The calibrated parameters for delayed Heston are

$$(V_0, \gamma, \theta^2, \delta, \rho, \alpha, \tau) = (0.0343, 3.9037, 10^{-8}, 0.808, -0.5057, 71.35, 0.7821),$$

and for Heston, they are

$$(V_0, \gamma, \theta^2, \delta, \rho) = (0.0328, 0.5829, 0.0256, 0.3672, -0.4824).$$

We notice that we cannot compare straightforwardly the parameters θ^2 of both models. Indeed, as mentioned above, the delayed Heston model has a time-dependent long-range variance $\tilde{\theta}_t^2$ which has been shifted away from its original value θ^2 because of the introduction of the delay τ. When $\tau = 0$, $\tilde{\theta}_t^2 = \theta^2$, but when $\tau > 0$, $\tilde{\theta}_t^2$ and θ^2 differ. Therefore, to be coherent, one should compare the Heston long-range variance $\theta^2 = 0.0256$ with the delayed Heston time-dependent long-range variance $\tilde{\theta}_t^2$. Below we give the value of $\tilde{\theta}_t^2$ for different maturities t (Table 6.1).

Table 6.1 Parameter $\tilde{\theta}_t^2$ for different maturities t

Maturity	$\tilde{\theta}_t^2$
1M	0.0325
2M	0.0322
3M	0.0319
6M	0.031
1Y	0.0294
2Y	0.0364
5Y	0.0184
10Y	0.0102

We remark that the short- and medium-term values (less than 2Y) of $\tilde{\theta}_t^2$ are similar to the value of θ^2 in the Heston model, but that for long maturities, the value of $\tilde{\theta}_t^2$ decreases significantly. Allowing this time-dependence of the long-range variance could be an explanation why the delayed Heston model outperforms the Heston model especially for long maturities (see the discussion below). Similarly, the Heston mean-reversion speed $\gamma = 0.58$ has to be compared with the delayed Heston adjusted mean-reversion speed γ_τ, which is given by (6.13) and is approximately equal to 0.12 on our calibration date. Focusing on the delay parameters α and τ, they were expected to be significantly nonzero because as we will see below, the delayed Heston model significantly outperforms the Heston model in terms of calibration error (and standard deviation of the calibration errors): if α and τ were close to 0, the calibration errors would have been approximately the same for both models, because again, the delayed Heston model reduces to the Heston model when the delay term vanishes, i.e. when $\tau = 0$ or $\alpha = 0$.

The calibration errors for all call options (expressed as the absolute value of the difference between market and model implied Black and Scholes volatilities, in bp) for the Heston model and our delayed Heston model are given below. The results show a 44% reduction of the average absolute calibration error (46bp for delayed Heston, 81bp for Heston). It is to be noted that we didn't use any weight matrix in our calibration procedure, i.e. the calibration aims at minimizing the sum of the (squares of the) errors of each call option, equally weighted. In practice, one might be willing to give more importance to ATM options, for instance, or options of a certain range of maturities. The optimization algorithm aims at minimizing the sum of the squares of the errors: in other words, it aims at minimizing the average absolute calibration error. For this reason, it might be the case that for some specific options (e.g. ATM 6M; see table below), the Heston model has a lower model error than the delayed Heston model. But the total calibration error for the delayed Heston model is always expected to be lower than for the Heston model (Table 6.2).

On our calibration date, the delayed Heston model seems to outperform the Heston model specifically for long maturities ($\geq 3Y$): if we consider only these options, the average absolute error is of 79bp for the Heston model and 33bp for the Delayed Heston model, which represents a 58% reduction of the calibration error. We can also note that for ATM options only, the improvement is significant too (43bp Vs. 92bp, i.e. an error reduction of 54%). For medium maturity options (6M to 2Y), the delayed Heston model still outperforms the Heston model but less significantly (53bp Vs. 75bp, i.e. an error reduction of 30%), and we have the same observation for very out-of-the-money options (10 delta call and put, 51bp vs. 79bp, i.e. an error reduction of 35%) (Table 6.3).

Another very interesting observation we can make is that the standard deviation of the calibration errors is much lower for the delayed Heston model compared to the Heston model (34bp vs. 52bp, which represents a 35% reduction of the standard deviation): in addition to improving the average absolute calibration error, it also improves the distance of the individual errors to the average error, which is highly appreciable in practice because it means that you won't face the case where some options are priced really poorly by the model, whereas some others are priced almost perfectly (Table 6.4).

Table 6.2 Heston absolute calibration error (in bp of the Black and Scholes volatility)

	ATM	25D Call	25D Put	10D Call	10D Put
1M	152	192	41	193	67
2M	114	139	15	136	81
3M	89	109	3	110	92
4M	48	61	17	67	101
6M	5	15	34	29	85
9M	59	42	63	2	85
1Y	107	83	102	31	96
1.5Y	141	116	111	42	73
2Y	166	137	127	54	68
3Y	145	124	77	52	0
4Y	96	95	18	37	66
5Y	29	47	52	7	138
7Y	39	10	112	28	186
10Y	100	67	168	58	225

Table 6.3 Delayed Heston absolute calibration error (in bp of the Black and Scholes volatility)

	ATM	25D Call	25D Put	10D Call	10D Put
1M	116	91	109	128	115
2M	44	24	59	54	88
3M	14	3	32	36	60
4M	18	28	1	5	29
6M	31	37	23	19	3
9M	45	45	56	37	57
1Y	51	47	82	50	104
1.5Y	29	30	79	49	129
2Y	24	23	83	47	139
3Y	11	9	29	30	90
4Y	41	28	14	17	38
5Y	76	55	59	5	16
7Y	71	49	58	1	14
10Y	26	8	18	47	24

Table 6.4 Standard deviation of the calibration errors in bp. The reduction of this error is indicated in brackets

	Delayed Heston	Heston
Standard deviation of the calibration errors (bp)	33.67 (35%)	52.12

In order to i) check that our calibration on 30 September 2011 was not an exception and ii) investigate the stability of the calibrated parameters, we performed calibrations on 11 additional dates evenly spaced around 30 September 2011, ranging from 19 September 2011 to 17 October 2011. We chose a one-month window because from the past experience of the authors in the financial industry, it can happen that the parameters are recalibrated by financial institutions every month only and not everyday (because it would be too time-consuming), and therefore the choice of a one-month window seems reasonable to investigate the stability of the parameters.

We summarize the findings in Tables 6.5 and 6.6. We find that the delayed Heston model always outperforms significantly the Heston model (average calibration error reduction varying from 29% to 56%) and that the delayed Heston model is performant especially for long maturities ($\geq$ 3Y, calibration error reduction varying from 40% to 66%) and ATM options (calibration error reduction varying from 42% to 67%). Finally, the standard deviation of the calibration errors is always reduced significantly by the delayed Heston model (reduction varying from 23% to 49%).

In order to investigate the stability of the model parameters, we present below the calibrated parameters for the Heston model and the delayed Heston model from 19 September 2011 to 17 October 2011 (Tables 6.7 and 6.8).

We can see that in average, the parameters stay relatively stable throughout this one-month time window. In this case, it would be reasonable to use the same parameters throughout the one-month time window as some financial institutions do (from the past experience of the authors in the financial industry). Of course, there are some periods of high volatility in which not recalibrating the model parameters often enough might lead to a significant mispricing of the call options by the model.

6.4 Pricing Variance and Volatility Swaps

In this section, we derive a closed formula for the Brockhaus and Long approximation of the volatility swap price using the change of time method introduced in Swishchuk (2004), as well as the price of the variance swap. Precisely, in Brockhaus and Long (2000), the following approximation was presented to compute the expected value of the square root of an almost surely non-negative random variable Z: $\mathbb{E}(\sqrt{Z}) \approx \sqrt{\mathbb{E}(Z)} - \frac{Var(Z)}{8\mathbb{E}(Z)^{\frac{3}{2}}}$. We denote $V_R := \frac{1}{T} \int_0^T V_s ds$ the realized variance on $[0, T]$.

We let $X_t(T) := \mathbb{E}_t^{\mathbb{Q}}(V_R)$ (resp. $Y_t(T) := \mathbb{E}_t^{\mathbb{Q}}(\sqrt{V_R})$) the price process of the floating leg of the variance swap (resp. volatility swap) of maturity T.

Theroem 1: The price process $X_t(T)$ of the floating leg of the variance swap of maturity T in the delayed Heston model (6.5)–(6.8) is given by

$$
\begin{aligned}
X_t(T) = \frac{1}{T} \int_0^t V_s ds &+ \frac{T-t}{T} \theta_\tau^2 + (V_t - \theta_\tau^2)\left(\frac{1 - e^{-\gamma(T-t)}}{\gamma T}\right) \\
&+ (V_0 - \theta_\tau^2)e^{-\gamma_\tau t}\left(\frac{1 - e^{-\gamma_\tau(T-t)}}{\gamma_\tau T} - \frac{1 - e^{-\gamma(T-t)}}{\gamma T}\right).
\end{aligned}
\tag{6.18}
$$

Table 6.5 Summary of the calibration error reductions

Date (2011)	Sep. 19	Sep. 21	Sep. 23	Sep. 27	Sep. 29	Oct. 3	Oct. 5	Oct. 7	Oct. 11	Oct. 13	Oct. 17
Total error reduction (%)	44	45	56	47	38	51	42	37	38	29	39
Long maturity error reduction (%)	58	63	65	55	53	66	55	51	48	40	59
ATM error reduction (%)	57	55	67	62	55	65	56	51	50	42	53

Table 6.6 Summary of the calibration errors st. dev. reductions

Date (2011)	Sep. 19	Sep. 21	Sep. 23	Sep. 27	Sep. 29	Oct. 3	Oct. 5	Oct. 7	Oct. 11	Oct. 13	Oct. 17
Calibration errors st. dev. reduction (%)	43	45	49	46	31	49	40	29	29	23	29

Table 6.7 Calibrated parameters for the delayed Heston model

Date (2011)	Sep. 19	Sep. 21	Sep. 23	Sep. 27	Sep. 29	Oct. 3	Oct. 5	Oct. 7	Oct. 11	Oct. 13	Oct. 17
V_0	0.0313	0.0337	0.0384	0.0354	0.0335	0.0368	0.0344	0.0295	0.0279	0.0271	0.0283
γ	3.99	3.72	3.82	3.72	4.52	3.47	3.86	3.71	3.13	3.08	3.39
θ^2	5 e-4	7 e-6	2e-4	1e-8	1e-8	1e-5	3e-4	2e-3	1e-3	5e-3	4e-3
δ	0.79	0.75	0.82	0.81	0.89	0.78	0.81	0.76	0.68	0.67	0.80
ρ	-0.51	-0.51	-0.52	-0.50	-0.49	-0.51	-0.51	-0.51	-0.51	-0.51	-0.49
α	82.2	77.5	64.5	166.7	124	66.6	76.5	90.2	125.1	83.7	77.3
τ	0.86	0.77	0.71	0.32	0.59	0.72	0.78	0.90	0.67	1.00	0.81

Table 6.8 Calibrated parameters for the Heston model

Date (2011)	Sep. 19	Sep. 21	Sep. 23	Sep. 27	Sep. 29	Oct. 3	Oct. 5	Oct. 7	Oct. 11	Oct. 13	Oct. 17
V_0	0.0298	0.0322	0.0369	0.0338	0.0311	0.0351	0.0326	0.0283	0.0271	0.0262	0.0269
γ	0.54	0.46	0.45	0.43	0.35	0.44	0.43	0.89	1.22	1.13	0.92
θ^2	0.0258	0.026	0.0258	0.0264	0.0276	0.0265	0.0275	0.0265	0.0249	0.0254	0.024
δ	0.34	0.33	0.35	0.35	0.32	0.34	0.34	0.41	0.45	0.43	0.39
ρ	-0.49	-0.49	-0.49	-0.48	-0.48	-0.50	-0.49	-0.50	-0.50	-0.49	-0.51

Proof: By definition, $X_t(T) = \mathbb{E}_t^{\mathbb{Q}}(\frac{1}{T}\int_0^T V_s ds) = \frac{1}{T}\int_0^t V_s ds + \frac{1}{T}\int_t^T \mathbb{E}_t^{\mathbb{Q}}(V_s) ds$. In the previous integral, the interchange between the expectation and the integral is justified by the use of Tonelli's theorem, as the variance process $(t, \omega) \to V_t(\omega)$ is a.e. non-negative and measurable. Let $s \geq t$. Then we have by (6.8) that $\mathbb{E}_t^{\mathbb{Q}}(V_s - V_t) = \mathbb{E}_t^{\mathbb{Q}}(V_s) - V_t = \int_t^s \gamma(\theta^2 - \mathbb{E}_t^{\mathbb{Q}}(V_u)) + \varepsilon_\tau(u) du + \mathbb{E}_t^{\mathbb{Q}}(\int_t^s \sqrt{V_u} dW_u^{\mathbb{Q}})$. Again, the interchange of the expectation and the integral $\mathbb{E}_t^{\mathbb{Q}}(\int_t^s \gamma(\theta^2 - V_u) + \varepsilon_\tau(u) du) = \int_t^s \gamma(\theta^2 - \mathbb{E}_t^{\mathbb{Q}}(V_u)) + \varepsilon_\tau(u) du$ is obtained the following way:

$$\mathbb{E}_t^{\mathbb{Q}}(\int_t^s \gamma(\theta^2 - V_u) + \varepsilon_\tau(u) du) = \int_t^s \gamma\theta^2 + \varepsilon_\tau(u) du - \gamma\mathbb{E}_t^{\mathbb{Q}}(\int_t^s V_u du). \quad (6.19)$$

Then again, by Tonelli's theorem we get $\mathbb{E}_t^{\mathbb{Q}}(\int_t^s V_u du) = \int_t^s \mathbb{E}_t^{\mathbb{Q}}(V_u) du$, which justifies the interchange.

Now, $(\sqrt{V_t})_{t \geq 0}$ is an adapted process (to our filtration $(\mathscr{F}_t)_{t \geq 0}$) s.t. $\mathbb{E}^{\mathbb{Q}}(\int_0^T V_u du) = \int_0^T \mathbb{E}^{\mathbb{Q}}(V_u) du < +\infty$ (by Tonelli's theorem); therefore $\int_0^t \sqrt{V_u} dW_u^{\mathbb{Q}}$ is a martingale and we have $\mathbb{E}_t^{\mathbb{Q}}(\int_t^s \sqrt{V_u} dW_u^{\mathbb{Q}}) = 0$. Therefore $\forall s \geq t \geq 0$; the function $s \to \mathbb{E}_t^{\mathbb{Q}}(V_s)$ is a solution of $y_s' = \gamma(\theta^2 - y_s) + \varepsilon_\tau(s)$ with initial condition $y_t = V_t$. This gives us $\mathbb{E}_t^{\mathbb{Q}}(V_s) = \theta_\tau^2 + (V_t - \theta_\tau^2)e^{-\gamma(s-t)} + (V_0 - \theta_\tau^2)e^{-\gamma_\tau t}(e^{-\gamma_\tau(s-t)} - e^{-\gamma(s-t)})$. Integrating the latter in the variable s via $\int_t^T \mathbb{E}_t^{\mathbb{Q}}(V_s) ds$ completes the proof.

Corollary 1: The price K_{var} of the variance swap of maturity T at initiation of the contract $t = 0$ in the delayed Heston model (6.5)–(6.8) is given by

$$K_{var} = \theta_\tau^2 + (V_0 - \theta_\tau^2)\frac{1 - e^{-\gamma_\tau T}}{\gamma_\tau T}. \quad (6.20)$$

Proof: By definition, $K_{var} = X_0(T)$.

Now, let

$$x_t := -(V_0 - \theta_\tau^2)e^{(\gamma - \gamma_\tau)t} + e^{\gamma t}(V_t - \theta_\tau^2). \quad (6.21)$$

Then by Ito's lemma, we get

$$dx_t = \delta e^{\gamma t}\sqrt{(x_t + (V_0 - \theta_\tau^2)e^{(\gamma - \gamma_\tau)t})e^{-\gamma t} + \theta_\tau^2} dW_t^{\mathbb{Q}}. \quad (6.22)$$

which is of the form $dx_t = f(t, x_t) dW_t^{\mathbb{Q}}$ with

$$f(t, x) := \delta e^{\gamma t}\sqrt{(x + (V_0 - \theta_\tau^2)e^{(\gamma - \gamma_\tau)t})e^{-\gamma t} + \theta_\tau^2}. \quad (6.23)$$

Indeed, since $x_t = g(t, V_t)$ with $g(t, x) := -(V_0 - \theta_\tau^2)e^{(\gamma - \gamma_\tau)t} + e^{\gamma t}(x - \theta_\tau^2)$, the multidimensional version of Ito's lemma reads

$$dx_t = dg(t, V_t) = g_t(t, V_t) dt + g_x(t, V_t) dV_t + \frac{1}{2}g_{xx}(t, V_t) d\langle V, V\rangle_t, \quad (6.24)$$

where $\langle V, V \rangle_t$ is the quadratic variation of the process $(V_t)_{t \geq 0}$ (e.g. see Karatzas and Shreve 1998, Theorem 3.6. of Section 3.3). Since $g_{xx}(t,x) = 0$, $g_t(t,V_t) = -(\gamma - \gamma_\tau)(V_0 - \theta_\tau^2)e^{(\gamma - \gamma_\tau)t} + \gamma e^{\gamma t}(V_t - \theta_\tau^2)$ and $g_x(t,V_t) = e^{\gamma t}$, we get, using (6.8), (6.12), and (6.14),

$$dx_t = g_t(t,V_t)dt + g_x(t,V_t)dV_t \tag{6.25}$$

$$= -(\gamma - \gamma_\tau)(V_0 - \theta_\tau^2)e^{(\gamma - \gamma_\tau)t}dt + \gamma e^{\gamma t}(V_t - \theta_\tau^2)dt + e^{\gamma t}dV_t \tag{6.26}$$

$$= -e^{\gamma t}(\varepsilon_\tau(t) - \gamma(\theta_\tau^2 - \theta^2))dt + \gamma e^{\gamma t}(V_t - \theta_\tau^2)dt \tag{6.27}$$

$$+ e^{\gamma t}\left[\gamma(\theta^2 - V_t) + \varepsilon_\tau(t)\right]dt + e^{\gamma t}\delta\sqrt{V_t}dW_t^{\mathbb{Q}} \tag{6.28}$$

$$= e^{\gamma t}\delta\sqrt{V_t}dW_t^{\mathbb{Q}}. \tag{6.29}$$

The fact that $V_t = (x_t + (V_0 - \theta_\tau^2)e^{(\gamma - \gamma_\tau)t})e^{-\gamma t} + \theta_\tau^2$ by definition of x_t (6.21) completes the proof.

Because $dx_t = f(t,x_t)dW_t^{\mathbb{Q}}$, the process $(x_t)_{t \geq 0}$ is a continuous local martingale, and even a true martingale since $\mathbb{E}^{\mathbb{Q}}(\int_0^T f^2(s,x_s)ds) = \int_0^T \mathbb{E}^{\mathbb{Q}}(f^2(s,x_s))ds < \infty$ (again, the interchange between expectation and integral follows from Tonelli's theorem). We can use the change of time method introduced in Swishchuk (2004) and we get $x_t = \tilde{W}_{T_t}$, where $\tilde{W}_t$ is a $\mathscr{F}_{\hat{T}_t}$−adapted $\mathbb{Q}$−Brownian motion, which is based on the fact that every continuous local martingale can be represented as a time-changed Brownian motion. The process $(T_t)_{t \geq 0}$ is a.e. increasing, non-negative, and $\mathscr{F}_t$ adapted and is called the change of time process. This process is also equal to the quadratic variation $\langle x \rangle_t$ of the (square integrable) continuous martingale x_t (see Karatzas and Shreve 1998, Section 3.2, Proposition 2.10.).

Expressions of T_t, $\hat{T}_t$ and $\tilde{W}_t$ are given by

$$T_t = \langle x \rangle_t = \int_0^t f^2(s,x_s)\,ds, \tag{6.30}$$

$$\tilde{W}_t = \int_0^{\hat{T}_t} f(s,x_s)dW_s^{\mathbb{Q}}, \tag{6.31}$$

$$\hat{T}_t = \int_0^t \frac{1}{f^2\left(\hat{T}_s, x_{\hat{T}_s}\right)}ds. \tag{6.32}$$

To see that $\hat{T}_t$ has the following form, observe that

$$\hat{T}_{T_t} = \int_0^{T_t} \frac{1}{f^2\left(\hat{T}_s, x_{\hat{T}_s}\right)}ds. \tag{6.33}$$

Now make the change of variable $s = T_u$, so that $ds = dT_u = f^2(u, x_u)\,du$. We get

$$\hat{T}_{T_t} = \int_0^t \frac{f^2(u, x_u)}{f^2\left(\hat{T}_{T_u}, x_{\hat{T}_{T_u}}\right)}\,du = \int_0^t \frac{f^2(u, x_u)}{f^2(u, x_u)}\,du = t. \tag{6.34}$$

This immediately yields

$$V_t = \theta_\tau^2 + (V_0 - \theta_\tau^2)e^{-\gamma_\tau t} + e^{-\gamma t}\tilde{W}_{T_t}. \tag{6.35}$$

Lemma 1: For $s, t \geq 0$, we have

$$\mathbb{E}_t^{\mathbb{Q}}(\tilde{W}_{T_s}) = \tilde{W}_{T_{t \wedge s}}, \tag{6.36}$$

and for $s, u \geq t$

$$\begin{aligned}
\mathbb{E}_t^{\mathbb{Q}}(\tilde{W}_{T_s}\tilde{W}_{T_u}) = x_t^2 + \delta^2 &\left[\theta_\tau^2\left(\frac{e^{2\gamma(s \wedge u)} - e^{2\gamma t}}{2\gamma}\right)\right. \\
&\left. + (V_0 - \theta_\tau^2)\left(\frac{e^{(2\gamma - \gamma_\tau)(s \wedge u)} - e^{(2\gamma - \gamma_\tau)t}}{2\gamma - \gamma_\tau}\right) + x_t\left(\frac{e^{\gamma(s \wedge u)} - e^{\gamma t}}{\gamma}\right)\right].
\end{aligned} \tag{6.37}$$

Proof: (6.36) comes from the fact that $x_t = \tilde{W}_{T_t}$ is a martingale. Let $s \geq u \geq t$. Then by iterated conditioning $\mathbb{E}_t^{\mathbb{Q}}(\tilde{W}_{T_s}\tilde{W}_{T_u}) = \mathbb{E}_t^{\mathbb{Q}}(\mathbb{E}_u^{\mathbb{Q}}(\tilde{W}_{T_s}\tilde{W}_{T_u})) = \mathbb{E}_t^{\mathbb{Q}}(\tilde{W}_{T_u}\mathbb{E}_u^{\mathbb{Q}}(\tilde{W}_{T_{i_s}})) = \mathbb{E}_t^{\mathbb{Q}}(\tilde{W}_{T_u}^2)$, because $x_t = \tilde{W}_{T_t}$ is a martingale. Now, by definition of the quadratic variation, $x_u^2 - \langle x \rangle_u$ is a martingale, and therefore $\mathbb{E}_t^{\mathbb{Q}}(\tilde{W}_{T_u}^2) = x_t^2 - \langle x \rangle_t + \mathbb{E}_t^{\mathbb{Q}}(\langle x \rangle_u) = x_t^2 - T_t + \mathbb{E}_t^{\mathbb{Q}}(T_u) = x_t^2 - T_t + T_t + \mathbb{E}_t^{\mathbb{Q}}(\int_t^u f^2(s, x_s)\,ds)$. We can again interchange expectation and integral by Tonelli's theorem. By definition of $f^2(s, x_s)$ (the latter is a linear function of x_s) and since x_t martingale, then we have (for $s \geq t$) $\mathbb{E}_t^{\mathbb{Q}}(f^2(s, x_s)) = f^2(s, x_t)$, and therefore $\mathbb{E}_t^{\mathbb{Q}}(\tilde{W}_{T_s}\tilde{W}_{T_u}) = x_t^2 + \int_t^u f^2(s, x_t)\,ds$. We use the fact that, by definition of f in (6.23),

$$f^2(s, x_t) = \delta^2 e^{2\gamma s}[(x_t + (V_0 - \theta_\tau^2)e^{(\gamma - \gamma_\tau)s})e^{-\gamma s} + \theta_\tau^2], \tag{6.38}$$

to integrate the latter expression with respect to s to complete the proof.

The following theorem gives the expression of the Brockhaus and Long approximation of the volatility swap floating leg price process $Y_t(T)$.

Theroem 2: The Brockhaus and Long approximation of the price process $Y_t(T)$ of the floating leg of the volatility swap of maturity T in the delayed Heston model (6.5)–(6.8) is given by

$$Y_t(T) \approx \sqrt{X_t(T)} - \frac{Var_t^{\mathbb{Q}}(V_R)}{8X_t(T)^{\frac{3}{2}}}, \tag{6.39}$$

where $X_t(T)$ is given by equation (6.18) of Theorem 1 and

$$Var_t^Q(V_R) = \frac{x_t \delta^2}{\gamma^3 T^2} \left[e^{-\gamma t}\left(1 - e^{-2\gamma(T-t)}\right) - 2(T-t)\gamma e^{-\gamma T}\right]$$

$$+ \frac{\delta^2}{2\gamma^3 T^2}\left[2\theta_\tau^2\gamma(T-t) + 2(V_0 - \theta_\tau^2)\frac{\gamma}{\gamma_\tau}e^{-\gamma_\tau t} + 4\theta_\tau^2 e^{-\gamma(T-t)} - \theta_\tau^2 e^{-2\gamma(T-t)} - 3\theta_\tau^2\right]$$

$$- \frac{\delta^2(V_0 - \theta_\tau^2)}{\gamma^2 T^2(\gamma_\tau^2 + 2\gamma^2 - 3\gamma\gamma_\tau)}\left[2(\gamma_\tau - 2\gamma)e^{-\gamma(T-t)-\gamma_\tau t}\right.$$

$$\left. + (\gamma - \gamma_\tau)e^{-2\gamma(T-t)-\gamma_\tau t} + 2\frac{\gamma^2}{\gamma_\tau}e^{-\gamma_\tau T}\right].$$

$$(6.40)$$

Proof: The (conditioned) Brockhaus and Long approximation gives us

$$Y_t(T) = \mathbb{E}_t^Q(\sqrt{V_R}) \approx \sqrt{\mathbb{E}_t^Q(V_R)} - \frac{Var_t^Q(V_R)}{8\mathbb{E}_t^Q(V_R)^{\frac{3}{2}}} = \sqrt{X_t(T)} - \frac{Var_t^Q(V_R)}{8X_t(T)^{\frac{3}{2}}}.$$

Furthermore,

$$Var_t^Q(V_R) = \mathbb{E}_t^Q((V_R - \mathbb{E}_t^Q(V_R))^2)$$

$$= \frac{1}{T^2}\mathbb{E}_t^Q\left(\left(\int_0^T (V_s - \mathbb{E}_t^Q(V_s))ds\right)^2\right). \qquad (6.41)$$

From (6.35) we have $V_t = \theta_\tau^2 + (V_0 - \theta_\tau^2)e^{-\gamma_\tau t} + e^{-\gamma t}\tilde{W}_{T_t}$, and since $\tilde{W}_{T_t}$ is a martingale, $V_s - \mathbb{E}_t^Q(V_s) = 0$ if $s \leq t$, and $V_s - \mathbb{E}_t^Q(V_s) = e^{-\gamma s}(\tilde{W}_{T_s} - x_t)$ if $s > t$.

Therefore,

$$Var_t^Q(V_R) = \frac{1}{T^2}\mathbb{E}_t^Q\left(\left(\int_t^T e^{-\gamma s}(\tilde{W}_{T_s} - x_t)ds\right)^2\right)$$

$$= \frac{1}{T^2}x_t^2\left(\int_t^T e^{-\gamma s}ds\right)^2 + \frac{1}{T^2}\mathbb{E}_t^Q\left(\left(\int_t^T e^{-\gamma s}\tilde{W}_{T_s}ds\right)^2\right) \qquad (6.42)$$

$$- \frac{2}{T^2}x_t\left(\int_t^T e^{-\gamma s}\mathbb{E}_t^Q(\tilde{W}_{T_s})ds\right)\left(\int_t^T e^{-\gamma s}ds\right).$$

The interchange of expectation and integral in the last equation is justified the following way: by definition of $\tilde{W}_{T_s} = x_s$ in (6.21), we get

$$\mathbb{E}_t^Q\left(\int_t^T e^{-\gamma s}\tilde{W}_{T_s}ds\right) = \mathbb{E}_t^Q\left(\int_t^T -(V_0 - \theta_\tau^2)e^{-\gamma_\tau s} + V_s - \theta_\tau^2 ds\right) \qquad (6.43)$$

$$= \int_t^T -(V_0 - \theta_\tau^2)e^{-\gamma_\tau s} - \theta_\tau^2 ds + \mathbb{E}_t^Q\left(\int_t^T V_s ds\right). \qquad (6.44)$$

We can interchange expectation and integral in the latter expression by Tonelli's theorem, which gives

$$\mathbb{E}_t^Q\left(\int_t^T e^{-\gamma s}\tilde{W}_{T_s}ds\right) = \int_t^T -(V_0-\theta_\tau^2)e^{-\gamma s} - \theta_\tau^2 ds + \int_t^T \mathbb{E}_t^Q(V_s)ds \quad (6.45)$$

$$= \int_t^T e^{-\gamma s}\mathbb{E}_t^Q(\tilde{W}_{T_s})ds. \quad (6.46)$$

Now we continue our computation to get

$$Var_t^Q(V_R) = -\frac{1}{T^2}x_t^2\left(\int_t^T e^{-\gamma s}ds\right)^2 + \frac{1}{T^2}\mathbb{E}_t^Q\left(\left(\int_t^T e^{-\gamma s}\tilde{W}_{T_s}ds\right)^2\right)$$

$$= \frac{1}{T^2}\int_t^T\int_t^T e^{-\gamma(s+u)}\mathbb{E}_t^Q(\tilde{W}_{T_s}\tilde{W}_{T_u})dsdu - \frac{1}{T^2}x_t^2e^{-2\gamma t}\left(\frac{1-e^{-\gamma(T-t)}}{\gamma}\right)^2. \quad (6.47)$$

The interchange expectation-integral

$$\mathbb{E}_t^Q\left(\int_t^T\int_t^T e^{-\gamma(s+u)}\tilde{W}_{T_s}\tilde{W}_{T_u}dsdu\right) = \int_t^T\int_t^T e^{-\gamma(s+u)}\mathbb{E}_t^Q(\tilde{W}_{T_s}\tilde{W}_{T_u})dsdu \quad (6.48)$$

is justified the same way as above, using the definition of $\tilde{W}_{T_t} = x_t$ in (6.21) together with Tonelli's theorem. Finally, we use equation (6.37) of Lemma 1 and integrate the expression with respect to s and u to complete the proof.

Corollary 2: The Brockhaus and Long approximation of the volatility swap price K_{vol} of maturity T at initiation of the contract $t = 0$ in the delayed Heston model (6.5)–(6.8) is given by

$$K_{vol} \approx \sqrt{K_{var}} - \frac{Var_0^Q(V_R)}{8K_{var}^{\frac{3}{2}}}, \quad (6.49)$$

where K_{var} is given by formula (6.20) of Corollary 1 and

$$Var_0^Q(V_R) = \frac{\delta^2 e^{-2\gamma T}}{2T^2\gamma^3}\left[\theta_\tau^2\left(2\gamma Te^{2\gamma T} + 4e^{\gamma T} - 3e^{2\gamma T} - 1\right) + \frac{\gamma}{2\gamma - \gamma_\tau}(V_0 - \theta_\tau^2)\right.$$

$$\left.\left(2e^{2\gamma T}\left(2\frac{\gamma}{\gamma_\tau} - 1\right) - 4\gamma e^{\gamma T}\left(\frac{e^{(\gamma-\gamma_\tau)T}-1}{\gamma - \gamma_\tau}\right) + 4e^{\gamma T}\left(1 - \frac{\gamma}{\gamma_\tau}e^{(\gamma-\gamma_\tau)T}\right) - 2\right)\right]. \quad (6.50)$$

We notice that letting $\tau \to 0$ (and therefore $\gamma_\tau \to \gamma$), we get the formula of Swishchuk (2004).

    Proof: We have by definition $K_{vol} = Y_0(T)$, therefore the result is obtained from equation (6.40) of Theorem 2.

6.5 Volatility Swap Hedging

In this section, we consider dynamic hedging of volatility swaps using variance swaps, as the latter are fairly liquid, easy-to-trade derivatives. In the spirit of Broadie and Jain (2008), we consider a portfolio containing at time t one unit of volatility swap and β_t units of variance swaps, both of maturity T. Therefore the value Π_t of the portfolio at time t is

$$\Pi_t = e^{-r(T-t)} \left[Y_t(T) - K_{vol} + \beta_t (X_t(T) - K_{var}) \right]. \tag{6.51}$$

The portfolio is self-financing; therefore,

$$d\Pi_t = r\Pi_t dt + e^{-r(T-t)} \left[dY_t(T) + \beta_t dX_t(T) \right]. \tag{6.52}$$

The price processes $X_t(T)$ and $Y_t(T)$ can be expressed, denoting $I_t := \int_0^t V_s ds$ the accumulated variance at time t (known at this time):

$$X_t(T) = \mathbb{E}_t^Q \left[\frac{1}{T} I_t + \frac{1}{T} \int_t^T V_s ds \right] = g(t, I_t, V_t), \tag{6.53}$$

$$Y_t(T) = \mathbb{E}_t^Q \left[\sqrt{\frac{1}{T} I_t + \frac{1}{T} \int_t^T V_s ds} \right] = h(t, I_t, V_t). \tag{6.54}$$

Remembering that $\tilde{\theta}_t^2 = \theta_\tau^2 + (V_0 - \theta_\tau^2) \frac{(\gamma - \gamma_t)}{\gamma} e^{-\gamma_t t}$ and noticing that $dI_t = V_t dt$, by Ito's lemma we get

$$dX_t(T) = \left[\frac{\partial g}{\partial t} + \frac{\partial g}{\partial I_t} V_t + \frac{\partial g}{\partial V_t} \gamma(\tilde{\theta}_t^2 - V_t) + \frac{1}{2} \frac{\partial^2 g}{\partial V_t^2} \delta^2 V_t \right] dt + \frac{\partial g}{\partial V_t} \delta \sqrt{V_t} dW_t^Q, \tag{6.55}$$

$$dY_t(T) = \left[\frac{\partial h}{\partial t} + \frac{\partial h}{\partial I_t} V_t + \frac{\partial h}{\partial V_t} \gamma(\tilde{\theta}_t^2 - V_t) + \frac{1}{2} \frac{\partial^2 h}{\partial V_t^2} \delta^2 V_t \right] dt + \frac{\partial h}{\partial V_t} \delta \sqrt{V_t} dW_t^Q. \tag{6.56}$$

As conditional expectations of cash flows at maturity of the contract, the price processes $X_t(T)$ and $Y_t(T)$ are by construction martingales, and therefore we should have

$$\frac{\partial g}{\partial t} + \frac{\partial g}{\partial I_t} V_t + \frac{\partial g}{\partial V_t} \gamma(\tilde{\theta}_t^2 - V_t) + \frac{1}{2} \frac{\partial^2 g}{\partial V_t^2} \delta^2 V_t = 0, \tag{6.57}$$

$$\frac{\partial h}{\partial t} + \frac{\partial h}{\partial I_t} V_t + \frac{\partial h}{\partial V_t} \gamma(\tilde{\theta}_t^2 - V_t) + \frac{1}{2} \frac{\partial^2 h}{\partial V_t^2} \delta^2 V_t = 0. \tag{6.58}$$

The second equation, combined with some appropriate boundary conditions, was used in Broadie and Jain (2008) to compute the value of the price process $Y_t(T)$, whereas we focus on its Brockhaus and Long approximation.

Therefore we get

$$dX_t(T) = \frac{\partial g}{\partial V_t} \delta \sqrt{V_t} dW_t^Q, \tag{6.59}$$

$$dY_t(T) = \frac{\partial h}{\partial V_t} \delta \sqrt{V_t} dW_t^Q. \tag{6.60}$$

and so

$$d\Pi_t = r\Pi_t dt + e^{-r(T-t)} \left[\frac{\partial h}{\partial V_t} \delta \sqrt{V_t} dW_t^Q + \beta_t \frac{\partial g}{\partial V_t} \delta \sqrt{V_t} dW_t^Q \right]. \tag{6.61}$$

In order to dynamically hedge a volatility swap of maturity T, one should therefore hold β_t units of variance swap of maturity T, with:

$$\beta_t = -\frac{\frac{\partial h}{\partial V_t}}{\frac{\partial g}{\partial V_t}} = -\frac{\frac{\partial Y_t(T)}{\partial V_t}}{\frac{\partial X_t(T)}{\partial V_t}}. \tag{6.62}$$

Remembering that $Var_0^Q(V_R)$ and K_{var} are given, respectively, in Corollaries 2 and 1, the initial hedge ratio β_0 is given by

$$\beta_0 = -\frac{\frac{\partial Y_0(T)}{\partial V_0}}{\frac{\partial X_0(T)}{\partial V_0}}, \tag{6.63}$$

$$\frac{\partial X_0(T)}{\partial V_0} = \frac{1 - e^{-\gamma_\tau T}}{\gamma_\tau T}, \tag{6.64}$$

$$\frac{\partial Y_0(T)}{\partial V_0} \approx \frac{\frac{\partial X_0(T)}{\partial V_0}}{2\sqrt{K_{var}}} - \frac{K_{var} \frac{\partial Var_0^Q(V_R)}{\partial V_0} - \frac{3}{2} \frac{\partial X_0(T)}{\partial V_0} Var_0^Q(V_R)}{8 K_{var}^{\frac{5}{2}}}, \tag{6.65}$$

$$\frac{\partial Var_0^Q(V_R)}{\partial V_0} = \frac{\delta^2 e^{-2\gamma T}}{T^2 \gamma^3} \frac{\gamma}{2\gamma - \gamma_\tau} \left[e^{2\gamma T} \left(2\frac{\gamma}{\gamma_\tau} - 1 \right) \right.$$

$$\left. -2\gamma e^{\gamma T} \left(\frac{e^{(\gamma - \gamma_\tau)T} - 1}{\gamma - \gamma_\tau} \right) + 2e^{\gamma T} \left(1 - \frac{\gamma}{\gamma_\tau} e^{(\gamma - \gamma_\tau)T} \right) - 1 \right]. \tag{6.66}$$

Remembering that $Var_t^Q(V_R)$ and $X_t(T)$ are given, respectively, in Theorems 2 and 1, the hedge ratio β_t for $t > 0$ is given by

$$\beta_t = -\frac{\frac{\partial Y_t(T)}{\partial V_t}}{\frac{\partial X_t(T)}{\partial V_t}}, \tag{6.67}$$

$$\frac{\partial X_t(T)}{\partial V_t} = \frac{1 - e^{-\gamma(T-t)}}{\gamma T}, \tag{6.68}$$

$$\frac{\partial Y_t(T)}{\partial V_t} \approx \frac{\frac{\partial X_t(T)}{\partial V_t}}{2\sqrt{X_t(T)}} - \frac{X_t(T)\frac{\partial Var_t^Q(V_R)}{\partial V_t} - \frac{3}{2}\frac{\partial X_t(T)}{\partial V_t}Var_t^Q(V_R)}{8X_t(T)^{\frac{5}{2}}}, \tag{6.69}$$

$$\frac{\partial Var_t^Q(V_R)}{\partial V_t} = \frac{\delta^2}{\gamma^3 T^2}\left[1 - e^{-2\gamma(T-t)} - 2(T-t)\gamma e^{-\gamma(T-t)}\right]. \tag{6.70}$$

We take the parameters that have been calibrated in Section 6.3 on 30 September 2011 and we plot the naive volatility swap strike $\sqrt{K_{var}}$ together with the adjusted volatility swap strike $\sqrt{K_{var}} - \frac{Var^Q(V_R)}{8K_{var}^{\frac{3}{2}}}$ along the maturity dimension, as well as the convexity adjustment $\frac{Var^Q(V_R)}{8K_{var}^{\frac{3}{2}}}$:

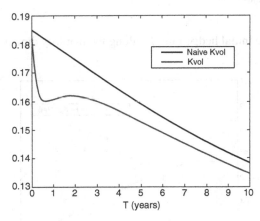

Fig. 6.1 Naive volatility swap strike vs. adjusted volatility swap strike

The naive volatility swap strike represents the initial fair value of the volatility swap contract obtained without taking into account the convexity adjustment $\frac{Var^Q(V_R)}{8K_{var}^{\frac{3}{2}}}$ linked to the Brockhaus and Long approximation, whereas the adjusted volatility swap strike represents this initial fair value when we do take into account the convexity adjustmentv (Fig. 6.1). The difference between the former and the latter is quantified by the convexity adjustment and is represented on the second graphic (Fig. 6.2). We see that neglecting the convexity adjustment leads to an overpricing of the volatility swap. On this example, the overpricing is especially significant for maturities less than 2Y, with a peak difference of more than 2% between the naive and adjusted strikes for maturities around 6M. The position of this local extremum (here, around 6M) is linked to the values of the calibrated parameters and therefore varies depending on the date we perform the calibration at.

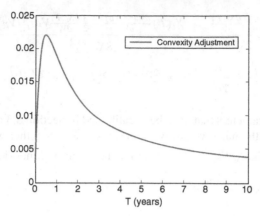

Fig. 6.2 Convexity adjustment

We also plot the initial hedge ratio β_0 along the maturity dimension (Fig. 6.3):

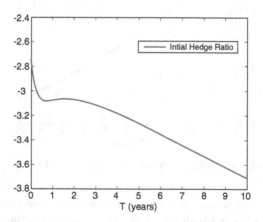

Fig. 6.3 Initial hedge ratio

This initial hedge ratio β_0 represents the quantity of variance swap contracts we need to buy (if $\beta_0 > 0$) or sell (if $\beta_0 < 0$) to hedge our position on one volatility swap contract of the same maturity. Of course, in order to cancel the risk, β_0 has to be negative if we buy a volatility swap contract and positive if we sell one. Here we have assumed that we hold a long position on a volatility swap contract, i.e. that we have bought one such contract. The plot tells us that for one volatility swap contract bought, we need to sell approximately three variance swap contracts of the same maturity (depending of the maturity of the contract) to hedge our position on the volatility swap, i.e. to cancel the risk inherent to our position. We say that we hold a short position on the variance swap contracts. The trend is that the higher the maturity of the volatility swap contract, the more variance swap contracts we need to sell in order to hedge our position. This was to be expected because for such pure volatility contracts, the longer the maturity, the higher the probability that the volatility varies significantly, i.e. the higher the risk.

6.6 Appendix: Semi-closed Formulas for Call Options in the Delayed Heston Model

From Kahl and Jäckel (2005), we get equations (6.71) to (6.74) for the price of a call option with maturity T and strike K in the time-dependent long-range variance Heston model:

$$C_0 = e^{-rT}\left[\frac{1}{2}(F-K) + \frac{1}{\pi}\int_0^\infty (Fh_1(u) - Kh_2(u))du\right], \qquad (6.71)$$

$$h_1(u) = \Re\left(\frac{e^{-iu\ln(K)}\varphi(u-i)}{iuF}\right), \qquad (6.72)$$

$$h_2(u) = \Re\left(\frac{e^{-iu\ln(K)}\varphi(u)}{iu}\right), \qquad (6.73)$$

with $F = S_0 e^{(r-q)T}$ and

$$\varphi(u) = e^{C(T,u)+V_0 D(T,u)+iu\ln(F)}. \qquad (6.74)$$

By Mikhailov and Noegel (2005), we have that $C(t,u)$ and $D(t,u)$ solve the following differential equations:

$$\frac{dC(t,u)}{dt} = \gamma\tilde{\theta}_t^2 D(t,u), \qquad (6.75)$$

$$\frac{dD(t,u)}{dt} - \frac{\delta^2}{2}D^2(t,u) + (\gamma - iu\rho\delta)D(t,u) + \frac{1}{2}(u^2 + iu) = 0, \qquad (6.76)$$

$$C(0,u) = D(0,u) = 0. \qquad (6.77)$$

The Riccati equation for $D(t,u)$ doesn't depend on $\tilde{\theta}_t^2$; therefore its solution is just the solution of the classical Heston model given in Kahl and Jäckel (2005):

$$D(t,u) = \frac{\gamma - ip\delta u + d}{\delta^2}\left[\frac{1-e^{dt}}{1-ge^{dt}}\right], \qquad (6.78)$$

$$g = \frac{\gamma - ip\delta u + d}{\gamma - ip\delta u - d}, \qquad (6.79)$$

$$d = \sqrt{(\gamma - ip\delta u)^2 + \delta^2(iu + u^2)}. \qquad (6.80)$$

Given $D(t,u)$ and the definition of $\tilde{\theta}_t^2$, we can compute $C(t,u)$ from (6.75) and (6.77):

$$C(t,u) = \gamma\theta_\tau^2 f(t,u) + (V_0 - \theta_\tau^2)(\gamma - \gamma_\tau)\int_0^t e^{-\gamma_\tau s}D(s,u)ds. \qquad (6.81)$$

where $f(t,u) = \int_0^t D(s,u)ds$ is given in Kahl and Jäckel (2005):

$$f(t,u) = \frac{1}{\delta^2}\left((\gamma - i\rho\delta u + d)t - 2\ln\left(\frac{1 - ge^{dt}}{1 - g}\right)\right). \tag{6.82}$$

Unfortunately, the integral $\int_0^t e^{-\gamma_{rs}}D(s,u)ds$ in (6.81) cannot be computed directly as $\int_0^t D(s,u)ds$. The logarithm in $f(t,u)$ can be handled as suggested in Kahl and Jäckel (2005), as well as the integration of the Heston integral, namely,

$$C_0 = e^{-rT}\int_0^1 y(x)dx, \tag{6.83}$$

$$y(x) = \frac{1}{2}(F - K) + \frac{Fh_1(-\frac{\ln(x)}{C_\infty}) - Kh_2(-\frac{\ln(x)}{C_\infty})}{x\pi C_\infty}, \tag{6.84}$$

where $C_\infty > 0$ is an integration constant.

The following limit conditions are given in Kahl and Jäckel (2005):

$$\lim_{x\to 0} y(x) = \frac{1}{2}(F - K), \tag{6.85}$$

$$\lim_{x\to 1} y(x) = \frac{1}{2}(F - K) + \frac{FH_1 - KH_2}{\pi C_\infty}, \tag{6.86}$$

$$H_j = \lim_{u\to 0} h_j(u) = \ln\left(\frac{F}{K}\right) + \tilde{c}_j(T) + V_0\tilde{d}_j(T), \tag{6.87}$$

where

$$\tilde{d}_1(t) = \Im\left(\frac{\partial D}{\partial u}(t, -i)\right), \tag{6.88}$$

$$\tilde{c}_1(t) = \Im\left(\frac{\partial C}{\partial u}(t, -i)\right), \tag{6.89}$$

$$\tilde{d}_2(t) = \Im\left(\frac{\partial D}{\partial u}(t, 0)\right), \tag{6.90}$$

$$\tilde{c}_2(t) = \Im\left(\frac{\partial C}{\partial u}(t, 0)\right). \tag{6.91}$$

Expressions for $\tilde{d}_1(t)$ and $\tilde{d}_2(t)$ are the same as in Kahl and Jäckel (2005) as $\bar{\theta}_t^2$ doesn't play any role in them. Given (6.75) and (6.77), we compute $\tilde{c}_1(T)$ and $\tilde{c}_2(T)$ in our time-dependent long-range variance Heston model by

$$\tilde{c}_j(T) = \gamma \int_0^T \bar{\theta}_t^2 \tilde{d}_j(t)dt. \tag{6.92}$$

After computing the integrals, we get the following:

If $\gamma - \rho\delta \neq 0$ and $\gamma - \rho\delta + \gamma_\tau \neq 0$,

$$\tilde{d}_1(T) = \frac{1 - e^{-(\gamma - \rho\delta)T}}{2(\gamma - \rho\delta)}, \tag{6.93}$$

$$\tilde{c}_1(T) = \gamma\theta_\tau^2 \frac{e^{-(\gamma - \rho\delta)T} - 1 + (\gamma - \rho\delta)T}{2(\gamma - \rho\delta)^2} \tag{6.94}$$

$$+ \frac{(V_0 - \theta_\tau^2)(\gamma - \gamma_\tau)}{2(\gamma - \rho\delta)} \left(-\frac{e^{-\gamma_\tau T} - 1}{\gamma_\tau} + \frac{e^{-(\gamma - \rho\delta + \gamma_\tau)T} - 1}{\gamma - \rho\delta + \gamma_\tau} \right). \tag{6.95}$$

If $\gamma - \rho\delta \neq 0$ and $\gamma - \rho\delta + \gamma_\tau = 0$,

$$\tilde{d}_1(T) = \frac{1 - e^{-(\gamma - \rho\delta)T}}{2(\gamma - \rho\delta)}, \tag{6.96}$$

$$\tilde{c}_1(T) = \gamma\theta_\tau^2 \frac{e^{-(\gamma - \rho\delta)T} - 1 + (\gamma - \rho\delta)T}{2(\gamma - \rho\delta)^2} \tag{6.97}$$

$$+ \frac{(V_0 - \theta_\tau^2)(\gamma - \gamma_\tau)}{2(\gamma - \rho\delta)} \left(-\frac{e^{-\gamma_\tau T} - 1}{\gamma_\tau} - T \right). \tag{6.98}$$

If $\gamma - \rho\delta = 0$,

$$\tilde{d}_1(T) = \frac{T}{2}, \tag{6.99}$$

$$\tilde{c}_1(T) = \gamma\theta_\tau^2 \frac{T^2}{4} + \frac{(V_0 - \theta_\tau^2)(\gamma - \gamma_\tau)}{2} \left(-\frac{Te^{-\gamma_\tau T}}{\gamma_\tau} + \frac{1 - e^{-\gamma_\tau T}}{\gamma_\tau^2} \right), \tag{6.100}$$

and

$$\tilde{d}_2(T) = \frac{e^{-\gamma T} - 1}{2\gamma}, \tag{6.101}$$

$$\tilde{c}_2(T) = \gamma\theta_\tau^2 \frac{1 - e^{-\gamma T} - \gamma T}{2\gamma^2} \tag{6.102}$$

$$+ \frac{(V_0 - \theta_\tau^2)(\gamma - \gamma_\tau)}{2\gamma} \left(-\frac{1 - e^{-\gamma_\tau T}}{\gamma_\tau} - \frac{e^{(-\gamma_\tau - \gamma)T} - 1}{\gamma_\tau + \gamma} \right). \tag{6.103}$$

References

Arriojas, M., Hu, Y., Mohammed, S-E., Pap, G. A Delayed Black and Scholes Formula *Stochastic Analysis and Applications* 25:2, 471–492, 2007.

Bell, D. R., Mohammed, S-E. The Malliavin calculus and stochastic delay equations. *J. Funct. Anal.* 99, no. 1, 75–99, 1991.

Bell, D. R., Mohammed, S-E. Smooth densities for degenerate stochastic delay equations with hereditary drift. *Ann. Probab.* 23, no. 4, 1875–1894, 1995.

Bollerslev, T. Generalized autoregressive conditional heteroscedasticity. *J. Economics* 31, 307–27, 1986.

Broadie, M. and Jain, A. The effect of jumps and discrete sampling on volatility and variance swaps. *Intern. J. Theor. Appl. Finance*, vol. 11, no. 8, pp. 761–797, 2008.

Broadie, M and Jain, A. Pricing and Hedging Volatility Derivatives. *Journal of Derivatives*, Vol.15, No.3, 7–24, 2008.

Brockhaus, O. and Long, D. Volatility swaps made simple. *Risk*, January, 92–96, 2000.

Carr, P. and Lee, R. Volatility Derivatives. *Annu. Rev, Financ. Econ.*, 1:1–21 (doi:10.1146/annurev.financial.050808.114304), 2009.

Filipovic, D. Affine Variance Swap Curve Models. *Seminar on Stochastic Analysis, Random Fields and Applications VII, Progress in Probability*, 67, 381–393, Springer Basel, 2013.

Hao, J. and Zhang, J.E. GARCH option pricing models, the CBOE VIX, and Variance Risk premium. *Journal of Financial Econometrics*, 11(3):556–580, 2013.

Heston, S. A closed-form solution for options with stochastic volatility with applications to bond and currency options. *Review of Financial Studies*, 6, 327–343, 1993.

Heston, S. and Nandi, S. Derivatives on volatility: some simple solutions based on observables. *Federal Reserve Bank of Atlanta*, working paper 2000–20, November, 2000.

Kahl, C. and Jäckel, P. Not-so-complex logarithms in the Heston model. *Wilmott Magazine*, September, pp. 94–103, 2005.

Karatzas, I. and Shreve, S.E., Brownian motion and Stochastic Calculus, 2nd Edition, *Springer Science+Business Media, LLC*, 1998.

Kazmerchuk, Y., Swishchuk, A. and Wu, J. A continuous-time GARCH model for stochastic volatility with delay. *Canadian Appl. Math. Quart.*, v. 13, No. 2, 2005.

Kazmerchuk, Y., Swishchuk, A. and Wu, J. The pricing of options for security markets with delayed response. *Mathematics and Computers in Simulation* 75 69–79, 2007.

Kind, P., Lipster, R. S., Runggaldier, W. J.. Diffusion approximation in past dependent models and applications to option pricing. *The Annals of Applied Probability* vol.1, no. 3, pp. 379–405, 1991.

Kruse, S. and Nögel, U. On the pricing of forward starting options in Heston's model on stochastic volatility. *Finance Stochast.* 9, 233–250, 2005.

Mikhailov, S. and Noegel, U. Heston's Stochastic Volatility Model Implementation, Calibration and Some Extensions. *Wilmott Magazine*, 1, pp. 74–79, 2005.

Swishchuk, A. and Li, X. Pricing variance swaps for stochastic volatilities with delay and jumps, *Intern. J. Stoch. Anal.*, vol. 2011 (2011), Article ID 435145, 2011.

Swishchuk, A. Modeling of variance and volatility swaps for financial markets with Stochastic volatilities. *Wilmott Magazine*, September issue, No. 2, pp. 64–72, 2004.

Swishchuk, A. Modeling and pricing of variance swaps for stochastic volatilities with delay. *Wilmott Magazine*, September, No. 19, pp. 63–73, 2005.

Swishchuk, A. Modeling and Pricing of Variance Swaps for Multi-Factor Stochastic Volatilities with Delay. *Canad. Appl. Math. Quart.*, Volume 14, Number 4, Winter 2006.

Swishchuk, A. Variance and volatility swaps in energy markets. *J. Energy Markets*, Volume 6/Number 1, Spring 2013 (33–49), 2013.

Swishchuk, A and Malenfant, K. Variance swap for local Levy based stochastic volatility with delay. *Intern. Rev. Appl. Fin. Iss. Econ. (IRAFIE)*, Vol. 3, No. 2, 2011, pp. 432–441, 2011.

Swishchuk, A. and Vadori, N. Smiling for the delayed volatility swaps. *Wilmott Magazine*, November 2014, 62–72.

Zhang, J.E. and Zhu, Y. Vix Futures. *The Journal of Futures Markets*, 2(6):521–531, 2006.

Chapter 7
CTM and the Explicit Option Pricing Formula for a Mean-Reverting Asset in Energy Markets

"Equations are more important to me, because politics is for the present, but an equation is something for eternity". —Albert Einstein.

Abstract In this chapter, we apply the CTM to get the explicit option pricing formula for a mean-reverting asset in energy markets.

7.1 Introduction

Some commodity prices, like oil and gas (see Chen and Forsyth (2006)), exhibit a mean reversion, unlike stock price. It means that over time, they tend to return to some long-term mean. In this chapter we consider a risky asset S_t following the mean-reverting stochastic process given by the following stochastic differential equation:

$$dS_t = a(L - S_t)dt + \sigma S_t dW_t,$$

where W is a standard Wiener process; $\sigma > 0$ is the volatility; the constant L is called the "long-term mean" of the process, to which it reverts to over time; and $a > 0$ measures the "strength" of the mean reversion.

This mean-reverting model is a one-factor version of the two-factor model made popular in the context of energy modelling by Pilipovic (1997). Black's model (1976) and Schwartz's model (1997) have become a standard approach to the problem of pricing options on commodities. These models have the advantage of mathematical convenience, in that they give rise to closed-form solutions for some types of options (see Wilmott 2000).

Bos et al. (2002) presented a method for evaluation of the price of a European option based on S_t, using a semi-spectral method. They did not have the convenience of a closed-form solution; however, they showed that values for certain types of options may nevertheless be found extremely efficiently. They used the following partial differential equation (see, e.g. Wilmott et al. 1995):

© The Author 2016
A. Swishchuk, *Change of Time Methods in Quantitative Finance*,
SpringerBriefs in Mathematics, DOI 10.1007/978-3-319-32408-1_7

$$C'_t + R(S,t)C'_S + \sigma^2 S^2 C''_{SS}/2 = rC$$

for option prices $C(S,t)$, where $R(S,t)$ depends only on S and t, and corresponds to the drift induced by the risk-neutral measure, and r is the risk-free interest rate. Simplifying this equation to singular diffusion equation, they were able to calculate numerically the solution.

The aim of this chapter is to obtain an explicit expression for a European option price, $C(S,t)$, based on S_t, using a change of time method (see Swishchuk 2007). This method was once applied by the author to price variance, volatility, covariance, and correlation swaps for the Heston model (see Swishchuk 2004).

7.2 The Explicit Option Pricing Formula for European Call Option for the Mean-Reverting Asset Model Under Physical Measure

Let $(\Omega, \mathscr{F}, \mathscr{F}_t, P)$ be a probability space with a sample space Ω, σ-algebra of Borel sets $\mathscr{F}$ and probability P. The filtration $\mathscr{F}_t$, $t \in [0,T]$, is the natural filtration of a standard Brownian motion W_t, $t \in [0,T]$, such that $\mathscr{F}_T = \mathscr{F}$.

Here, we consider a risky asset S_t following the mean-reverting stochastic process given by the following stochastic differential equation:

$$dS_t = a(L - S_t)dt + \sigma S_t dW_t, \qquad (7.1)$$

where W_t is an $\mathscr{F}_t$-measurable one-dimensional standard Wiener process; $\sigma > 0$ is the volatility; constant L is called the "long-term mean" of the process, to which it reverts to over time; and $a > 0$ measures the "strength" of the mean reversion.

In this section, we are going to obtain an explicit expression for a European option price, $C(S,t)$, based on S_t, using a change of time method and a physical measure (see Swishchuk (2008)).

7.2.1 Explicit Solution of MRAM

Let

$$V_t := e^{at}(S_t - L). \qquad (7.2)$$

Then, from (7.2) and (7.1), we obtain

$$dV_t = ae^{at}(S_t - L)dt + e^{at}dS_t = \sigma(V_t + e^{at}L)dW_t. \qquad (7.3)$$

Using a change of time approach in the equation (7.3) (see Ikeda and Watanabe 1981 or Elliott 1982), we obtain the following solution of the equation (7.3):

$$V_t = S_0 - L + \tilde{W}(\hat{T}_t),$$

or (see (7.2)),

$$S_t = e^{-at}[S_0 - L + \tilde{W}(\hat{T}_t)] + L, \qquad (7.4)$$

where $\tilde{W}(t)$ is an $\mathscr{F}_t$-measurable standard one-dimensional Wiener process and $\hat{T}_t$ is an inverse function to T_t :

$$T_t = \sigma^{-2} \int_0^t (S_0 - L + \tilde{W}(s) + e^{aT_s}L)^{-2}ds. \qquad (7.5)$$

We note that

$$\hat{T}_t = \sigma^2 \int_0^t (S_0 - L + \tilde{W}(\hat{T}_t) + e^{as}L)^2 ds, \qquad (7.6)$$

which follows from (7.5) and from the following transformations:

$$dT_t = \sigma^{-2}(S_0 - L + \tilde{W}(t) + e^{aT_t}L)^{-2}dt \Rightarrow \sigma^2(S_0 - L + \tilde{W}(t) + e^{aT_t}L)^2 d\phi_t = dt \Rightarrow$$

$$t = \sigma^2 \int_0^t (S_0 - L + \tilde{W}(s) + e^{a\phi_s}L)^2 d\phi_s \Rightarrow$$
$$\hat{T}_t = \sigma^2 \int_0^{\hat{T}_t} (S_0 - L + \tilde{W}(s) + e^{aT_s}L)^2 dT_s$$
$$= \sigma^2 \int_0^t (S_0 - L + \tilde{W}(\hat{T}_s) + e^{as}L)^2 ds.$$

7.2.2 Some Properties of the Process $\tilde{W}(\hat{T}_t)$

We note that process $\tilde{W}(\hat{T}_t)$ is $\tilde{\mathscr{F}}_t := \mathscr{F}_{\hat{T}_t}$-measurable and is $\tilde{\mathscr{F}}_t$-martingale.
 Then

$$E\tilde{W}(\hat{T}_t) = 0. \qquad (7.7)$$

Let us calculate the second moment of $\tilde{W}(\hat{T}_t)$ (see (7.6)):

$$\begin{aligned}
E\tilde{W}^2(\hat{T}_t) &= E < \tilde{W}(\hat{T}_t) >= E\hat{T}_t \\
&= \sigma^2 \int_0^t E(S_0 - L + \tilde{W}(\hat{T}_s) + e^{as}L)^2 ds \\
&= \sigma^2[(S_0 - L)^2 t + \tfrac{2L(S_0-L)(e^{at}-1)}{a} + \tfrac{L^2(e^{2at}-1)}{2a} \\
&\quad + \int_0^t E\tilde{W}^2(\hat{T}_s)ds].
\end{aligned} \qquad (7.8)$$

From (7.8), by solving this linear ordinary nonhomogeneous differential equation with respect to $E\tilde{W}^2(\hat{T}_t)$,

$$\frac{dE\tilde{W}^2(\hat{T}_t)}{dt} = \sigma^2[(S_0 - L)^2 + 2L(S_0 - L)e^{at} + L^2 e^{2at} + E\tilde{W}^2(\hat{T}_t)],$$

we obtain

$$E\tilde{W}^2(\hat{T}_t) = \sigma^2[(S_0 - L)^2 \frac{e^{\sigma^2 t} - 1}{\sigma^2} + \frac{2L(S_0 - L)(e^{at} - e^{\sigma^2 t})}{a - \sigma^2} + \frac{L^2(e^{2at} - e^{\sigma^2 t})}{2a - \sigma^2}].$$
$$(7.9)$$

7.2.3 The Explicit Expression for the Process $\tilde{W}(\hat{T}_t)$

It turns out that we can find the explicit expression for the process $\tilde{W}(\hat{T}_t)$.
From the following expression (see Section 3.1),

$$V_t = S_0 - L + \tilde{W}(\hat{T}_t),$$

we have the following relationship between $W(t)$ and $\tilde{W}(\hat{T}_t)$:

$$d\tilde{W}(\hat{T}_t) = \sigma \int_0^t [S(0) - L + Le^{at} + \tilde{W}(\hat{T}_s)] dW(t).$$

It is a linear SDE with respect to $\tilde{W}(\hat{T}_t)$ and we can solve it explicitly. The solution has the following look:

$$\tilde{W}(\hat{T}_t) = S(0)(e^{\sigma W(t) - \frac{\sigma^2 t}{2}} - 1) + L(1 - e^{at}) + aLe^{\sigma W(t) - \frac{\sigma^2 t}{2}} \int_0^t e^{as} e^{-\sigma W(s) + \frac{\sigma^2 s}{2}} ds.$$
$$(7.10)$$

It is easy to see from (7.10) that $\tilde{W}(\hat{T}_t)$ can be presented in the form of a linear combination of two zero-mean martingales $m_1(t)$ and $m_2(t)$:

$$\tilde{W}(\hat{T}_t) = m_1(t) + Lm_2(t),$$

where

$$m_1(t) := S(0)(e^{\sigma W(t) - \frac{\sigma^2 t}{2}} - 1)$$

and

$$m_2(t) = (1 - e^{at}) + ae^{\sigma W(t) - \frac{\sigma^2 t}{2}} \int_0^t e^{as} e^{-\sigma W(s) + \frac{\sigma^2 s}{2}} ds.$$

Indeed, process $\tilde{W}(\hat{T}_t)$ is a martingale (see Section 3.2). It is also well known that process $e^{\sigma W(t) - \frac{\sigma^2 t}{2}}$, and, hence, process $m_1(t)$ is a martingale. Then the process $m_2(t)$, as the difference between two martingales, is also martingale. In this way, we have

$$Em_1(t) = 0,$$

since

$$Ee^{\sigma W(t) - \frac{\sigma^2 t}{2}} = 1.$$

As for $m_2(t)$, we have

$$Em_2(t) = 0,$$

since from Itô's formula, we have

$$d(ae^{\sigma W(t)-\frac{\sigma^2 t}{2}} \int_0^t e^{as} e^{-\sigma W(s)+\frac{\sigma^2 s}{2}} ds) = a\sigma e^{\sigma W(t)-\frac{\sigma^2 t}{2}} \int_0^t e^{as} e^{-\sigma W(s)+\frac{\sigma^2 s}{2}} ds\, dW(t)$$
$$+ ae^{\sigma W(t)-\frac{\sigma^2 t}{2}} e^{at} e^{-\sigma W(t)+\frac{\sigma^2 t}{2}} dt$$
$$= a\sigma e^{\sigma W(t)-\frac{\sigma^2 t}{2}} \int_0^t e^{as} e^{-\sigma W(s)+\frac{\sigma^2 s}{2}} ds\, dW(t)$$
$$+ ae^{at} dt,$$

and, hence,

$$E ae^{\sigma W(t)-\frac{\sigma^2 t}{2}} \int_0^t e^{as} e^{-\sigma W(s)+\frac{\sigma^2 s}{2}} ds = e^{at} - 1.$$

It is interesting to see that in the last expression, the first moment for

$$\eta(t) := ae^{\sigma W(t)-\frac{\sigma^2 t}{2}} \int_0^t e^{as} e^{-\sigma W(s)+\frac{\sigma^2 s}{2}} ds,$$

does not depend on σ.

It is true only for the first moment but not for all the other moments of the process $\eta(t) = ae^{\sigma W(t)-\frac{\sigma^2 t}{2}} \int_0^t e^{as} e^{-\sigma W(s)+\frac{\sigma^2 s}{2}} ds$.

Indeed, using Itô's formula for $\eta^n(t)$, we obtain

$$d\eta^n(t) = na^n \sigma e^{n\sigma W(t)-\frac{n\sigma^2 t}{2}} (\int_0^t e^{as} e^{-\sigma W(s)+\frac{\sigma^2 s}{2}} ds)^n dW(t) + an(\eta_2(t))^{n-1} e^{at} dt$$
$$+ \frac{1}{2}n(n-1)\sigma^2 (\eta(t))^n dt,$$

and

$$dE\eta^n(t) = nae^{at} E\eta^{n-1}(t)dt + \frac{1}{2}n(n-1)\sigma^2 (\eta(t))^n dt, \quad n \geq 1.$$

This is a recursive equation with the initial function $(n = 1)$ $E\eta(t) = e^{at} - 1$. Thus, the expression for $E\eta^n(t)$ can be found for any $n \geq 2$.

7.2.4 Some Properties of Mean-Reverting Asset S_t

From (7.4) we obtain the mean value of the first moment for the mean-reverting asset S_t:

$$ES_t = e^{-at}[S_0 - L] + L.$$

It means that $ES_t \to L$ when $t \to +\infty$.

Using formulae (7.4) and (7.9), we can calculate the second moment of S_t:

$$ES_t^2 = (e^{-at}(S_0 - L) + L)^2$$
$$+ \sigma^2 e^{-2at}[(S_0 - L)^2 \frac{e^{\sigma^2 t}-1}{\sigma^2} + \frac{2L(S_0-L)(e^{at}-e^{\sigma^2 t})}{a-\sigma^2} + \frac{L^2(e^{2at}-e^{\sigma^2 t})}{2a-\sigma^2}].$$

Combining the first and the second moments, we have the variance of S_t :

$$Var(S_t) = ES_t^2 - (ES_t)^2$$
$$= \sigma^2 e^{-2at}[(S_0 - L)^2 \frac{e^{\sigma^2 t}-1}{\sigma^2} + \frac{2L(S_0-L)(e^{at}-e^{\sigma^2 t})}{a-\sigma^2} + \frac{L^2(e^{2at}-e^{\sigma^2 t})}{2a-\sigma^2}].$$

From the expression for $\tilde{W}(\hat{T}_t)$ (see (7.10)) and for $S(t)$ in (7.4), we can find the explicit expression for $S(t)$ through $W(t)$:

$$S(t) = e^{-at}[S_0 - L + \tilde{W}(\hat{T}_t)] + L$$
$$= e^{-at}[S_0 - L + m_1(t) + Lm_2(t)] + L \qquad (7.11)$$
$$= S(0)e^{-at}e^{\sigma W(t)-\frac{\sigma^2 t}{2}} + aLe^{-at}e^{\sigma W(t)-\frac{\sigma^2 t}{2}}\int_0^t e^{as}e^{-\sigma W(s)+\frac{\sigma^2 s}{2}}ds,$$

where $m_1(t)$ and $m_2(t)$ are defined as in Section 3.3.

7.2.5 The Explicit Option Pricing Formula for a European Call Option for the Mean-Reverting Asset Model Under Physical Measure

The payoff function f_T for European call option equals

$$f_T = (S_T - K)^+ := \max(S_T - K, 0),$$

where S_T is an asset price defined in (7.4), T is an expiration time (maturity), and K is a strike price.

In this way (see (7.11)),

$$f_T = [e^{-aT}(S_0 - L + \tilde{W}(\hat{T}_T)) + L - K]^+$$
$$= [S(0)e^{-aT}e^{\sigma W(T)-\frac{\sigma^2 T}{2}} + aLe^{-aT}e^{\sigma W(T)-\frac{\sigma^2 T}{2}}\int_0^T e^{as}e^{-\sigma W(s)+\frac{\sigma^2 s}{2}}ds - K]^+.$$

To find the option pricing formula, we need to calculate

$$C_T = e^{-rT}Ef_T$$
$$= e^{-rT}E[e^{-aT}(S_0 - L + \tilde{W}(\hat{T}_T)) + L - K]^+$$
$$= \frac{1}{\sqrt{2\pi}}e^{-rT}\int_{-\infty}^{+\infty} \max[S(0)e^{-aT}e^{\sigma y\sqrt{T}-\frac{\sigma^2 T}{2}} \qquad (7.12)$$
$$+ aLe^{-aT}e^{\sigma y\sqrt{T}-\frac{\sigma^2 T}{2}}\int_0^T e^{as}e^{-\sigma y\sqrt{s}+\frac{\sigma^2 s}{2}}ds - K, 0]e^{-\frac{y^2}{2}}dy.$$

Let y_0 be a solution of the following equation:

$$S(0) \times e^{-aT}e^{\sigma y_0\sqrt{T}-\frac{\sigma^2 T}{2}}$$
$$+ aLe^{-aT}e^{\sigma y_0\sqrt{T}-\frac{\sigma^2 T}{2}}\int_0^T e^{as}e^{-\sigma y_0\sqrt{s}+\frac{\sigma^2 s}{2}}ds = K \qquad (7.13)$$

or

$$y_0 = \frac{\ln(\frac{K}{S(0)}) + (\frac{\sigma^2}{2} + a)T}{\sigma\sqrt{T}}$$

$$- \frac{\ln(1 + \frac{aL}{S(0)} \int_0^T e^{as} e^{-\sigma y_0 \sqrt{s} + \frac{\sigma^2 s}{2}} ds)}{\sigma\sqrt{T}} \tag{7.14}$$

From (7.12)–(7.13), we have

$$
\begin{aligned}
C_T &= \frac{1}{\sqrt{2\pi}} e^{-rT} \int_{-\infty}^{+\infty} \max[S(0) e^{-aT} e^{\sigma y \sqrt{T} - \frac{\sigma^2 T}{2}} \\
&\quad + aLe^{-aT} e^{\sigma y \sqrt{T} - \frac{\sigma^2 T}{2}} \int_0^T e^{as} e^{-\sigma y \sqrt{s} + \frac{\sigma^2 s}{2}} ds - K, 0] e^{-\frac{y^2}{2}} dy \\
&= \frac{1}{\sqrt{2\pi}} e^{-rT} \int_{y_0}^{+\infty} [S(0) e^{-aT} e^{\sigma y \sqrt{T} - \frac{\sigma^2 T}{2}} \\
&\quad + aLe^{-aT} e^{\sigma y \sqrt{T} - \frac{\sigma^2 T}{2}} \int_0^T e^{as} e^{-\sigma y \sqrt{s} + \frac{\sigma^2 s}{2}} ds - K] e^{-\frac{y^2}{2}} dy \\
&= \frac{1}{\sqrt{2\pi}} e^{-rT} \int_{y_0}^{+\infty} [S(0) e^{-aT} e^{\sigma y \sqrt{T} - \frac{\sigma^2 T}{2}} e^{-\frac{y^2}{2}} dy - e^{-rT} K[1 - \Phi(y_0)] \\
&\quad + Le^{-(r+a)T} \frac{1}{\sqrt{2\pi}} \int_{y_0}^{+\infty} (ae^{\sigma y \sqrt{T} - \frac{\sigma^2 T}{2}} \int_0^T e^{as} e^{-\sigma y \sqrt{s} + \frac{\sigma^2 s}{2}} ds) e^{-\frac{y^2}{2}} dy \\
&= BS(T) + A(T),
\end{aligned}
\tag{7.15}
$$

where

$$BS(T) := \frac{1}{\sqrt{2\pi}} e^{-rT} \int_{y_0}^{+\infty} [S(0) e^{-aT} e^{\sigma y \sqrt{T} - \frac{\sigma^2 T}{2}} e^{-\frac{y^2}{2}} dy - e^{-rT} K[1 - \Phi(y_0)], \tag{7.16}$$

$$A(T) := Le^{-(r+a)T} \tag{7.17}$$
$$\times \frac{1}{\sqrt{2\pi}} \int_{y_0}^{+\infty} (ae^{\sigma y \sqrt{T} - \frac{\sigma^2 T}{2}} \int_0^T e^{as} e^{-\sigma y \sqrt{s} + \frac{\sigma^2 s}{2}} ds) e^{-\frac{y^2}{2}} dy,$$

and

$$\Phi(x) = \frac{1}{\sqrt{2\pi}} \int_{-\infty}^{x} e^{-\frac{y^2}{2}} dy. \tag{7.18}$$

After the calculation of $BS(T)$, we obtain

$$BS(T) = e^{-(r+a)T} S(0) \Phi(y_+) - e^{-rT} K\Phi(y_-), \tag{7.19}$$

where

$$y_+ := \sigma\sqrt{T} - y_0 \quad and \quad y_- := -y_0 \tag{7.20}$$

and y_0 is defined in (7.14).

Consider $A(T)$ in (7.17).

Let $F_T(dz)$ be a distribution function for the process

$$\eta(T) = ae^{\sigma W(T) - \frac{\sigma^2 T}{2}} \int_0^T e^{as} e^{-\sigma W(s) + \frac{\sigma^2 s}{2}} ds,$$

which is a part of the integrand in (7.17).

As Yor (1992) and Yor and Matsumoto (2005) mentioned, there is still no closed-form probability density function for a time integral of an exponential Brownian motion, while the best result is a function with a double integral.

We can use Yor's result (Yor 1992) to get $F_T(dz)$ above. Using the scaling property of the Wiener process and change of variables, we can rewrite our expression for $S(t)$ in (7.11) in the following way:

$$S(T) = S(0)e^{-2B^v_{T_0}} + \frac{4}{\sigma^2}aLe^{-2B^v_{T_0}}A^v_{T_0},$$

where $T_0 = \frac{\sigma^2}{4}T, v = \frac{2}{\sigma^2}a+1, B_t = -\frac{\sigma}{2}W(\frac{4}{\sigma^2}t), B^v_{T_0} = vT_0 + B_{T_0}, A^v_{T_0} = \int_0^{T_0} e^{2B^v_s}ds.$

Also, the process $\eta(T)$ may be presented in the following way using these transformations:

$$\eta(T) = \frac{4ae^{-aT}}{\sigma^2}e^{-2B_{\frac{\sigma^2 T}{4}}}A_{\frac{\sigma^2 T}{4}}.$$

For completeness, we state here one result for the joint probability density function of $A^v_{T_0}$ and $B^v_{T_0}$ obtained by Yor (1992).

Theroem 4.3.-1 (Yor (1992)). *The joint probability density function of $A^v_{T_0}$ and $B^v_{T_0}$ satisfies*

$$P(A^v_{T_0} \in du, B^v_{T_0} \in dx) = e^{vx-v^2t/2}\exp{-\frac{1+e^{2x}}{2u}}\theta(\frac{e^x}{u},t)\frac{dxdu}{u},$$

where $t > 0, u > 0, x \in R$, and

$$\theta(r,t) = \frac{r}{(2\pi^3 t)^{1/2}}e^{\frac{\pi^2}{2t}}\int_0^{+\infty} e^{-\frac{s^2}{2t}-r\cosh s}\sinh(s)\sin(\frac{\pi s}{t})ds.$$

Using this result we can write the distribution function for $\eta(T)$ in the following way:

$$\begin{aligned}
P(\eta(T) \leq u) &= P(\frac{4ae^{-aT}}{\sigma^2}e^{-2B_{\frac{\sigma^2 T}{4}}}A_{\frac{\sigma^2 T}{4}} \leq u) \\
&= P(e^{-2B_{\frac{\sigma^2 T}{4}}}A_{\frac{\sigma^2 T}{4}} \leq \frac{\sigma^2 e^{aT}}{4a}u) \\
&= F_T(u).
\end{aligned} \quad (7.21)$$

In this way, $A(T)$ in (7.17) may be presented in the following way:

$$A(T) = Le^{-(r+a)T}\int_{y_0}^{+\infty} zF_T(dz).$$

After the calculation of $A(T)$, we obtain the following expression for $A(T)$:

$$A(T) = Le^{-(r+a)T}[(e^{aT} - 1) - \int_0^{y_0} zF_T(dz)],$$

since $E\eta(T) = e^{aT} - 1$.

Finally, summarizing (7.12)–(7.21), we have obtained the following theorem:

Theorem 3.1. *The option pricing formula for a European call option for the mean-reverting asset under physical measure has the following form:*

$$C_T = e^{-(r+a)T} S(0)\Phi(y_+) - e^{-rT} K\Phi(y_-) \\ + Le^{-(r+a)T}[(e^{aT} - 1) - \int_0^{y_0} z F_T(dz)], \tag{7.22}$$

where y_0 is defined in (7.14), y_+ and y_- in (7.20), $\Phi(y)$ in (7.18), and $F_T(dz)$ is a distribution function in (7.21).

Remark. From (7.21)–(7.22), we find that the European call option price C_T for the mean-reverting asset lies between the following boundaries:

$$BS(T) \le C_T \le BS(T) + Le^{-(r+a)T}[e^{aT} - 1],$$

and (see (7.19))

$$e^{-(r+a)T} S(0)\Phi(y_+) - e^{-rT} K\Phi(y_-) \le C_T \\ \le e^{-(r+a)T} S(0)\Phi(y_+) \\ - e^{-rT} K\Phi(y_-) + Le^{-(r+a)T}[e^{aT} - 1].$$

7.3 The Mean-Reverting Risk-Neutral Asset Model (MRRNAM)

Consider our model (7.1):

$$dS_t = a(L - S_t)dt + \sigma S_t dW_t. \tag{7.23}$$

We want to find a probability P^* equivalent to P, under which the process $e^{-rt}S_t$ is a martingale, where $r > 0$ is a constant interest rate. The hypothesis we made on the filtration $(\mathscr{F}_t)_{t\in[0,T]}$ allows us to express the density of the probability P^* with respect to P. We denote this density by L_T.

It is well known (see Lamperton and Lapeyre 1996, Proposition 6.1.1, p. 123) that there is an adopted process $(q(t))_{t\in[0,T]}$ such that, for all $t \in [0,T]$,

$$L_t = \exp[\int_0^t q(s)dW_s - \frac{1}{2}\int_0^t q^2(s)ds] \quad a.s.$$

In this case,

$$\frac{dP^*}{dP} = \exp[\int_0^T q(s)dW_s - \frac{1}{2}\int_0^T q^2(s)ds] = L_T.$$

In our case, with model (7.17), the process $q(t)$ is equal to

$$q(t) = -\lambda S_t, \tag{7.24}$$

where λ is the *market price of risk* and $\lambda \in R$. Hence, for our model,

$$L_T = \exp[-\lambda \int_0^T S(u)dW_u - \frac{1}{2}\lambda \int_0^T S^2(u)du].$$

Under probability P^*, the process (W_t^*) defined by

$$W_t^* := W_t + \lambda \int_0^t S(u)du \qquad (7.25)$$

is a standard Brownian motion (Girsanov theorem) (see Elliott and Kopp 1999).

Therefore, in a risk-neutral world, our model (7.23) has the following form:

$$dS_t = (aL - (a+\lambda\sigma)S_t)dt + \sigma S_t dW_t^*,$$

or, equivalently,

$$dS_t = a^*(L^* - S_t)dt + \sigma S_t dW_t^*, \qquad (7.26)$$

where

$$a^* := a+\lambda\sigma, \quad L^* := \frac{aL}{a+\lambda\sigma}, \qquad (7.27)$$

and W_t^* is defined in (7.25).

Now, we have the same model in (7.26) as in (7.1), and we are going to apply our change of time method to this model (7.26) to obtain the explicit option pricing formula.

7.4 The Explicit Option Pricing Formula for a European Call Option for the Mean-Reverting Risk-Neutral Asset Model

In this section, we are going to obtain the explicit option pricing formula for the European call option under risk-neutral measure P^*, using the same arguments as in Sections 7.3, where instead of a and L, we are going to take a^* and L^*:

$$a \to a^* := a+\lambda\sigma, \quad L \to L^* := \frac{aL}{a+\lambda\sigma},$$

where λ is a *market price of risk* (see section 7.3).

7.4.1 The Explicit Solution for the Mean-Reverting Risk-Neutral Asset Model

Applying (7.2)–(7.6) to our model (7.26), we obtain the following explicit solution for our risk-neutral model (7.26):

$$S_t = e^{-a^*t}[S_0 - L^* + \tilde{W}^*((\hat{T}_t^*)] + L, \qquad (7.28)$$

where $\tilde{W}^*(t)$ is an $\mathscr{F}_t$-measurable standard one-dimensional Wiener process under measure P^* and $(\hat{T}_t^*$ is an inverse function to T_t^* :

$$T_t^* = \sigma^{-2} \int_0^t (S_0 - L^* + \tilde{W}^*(s) + e^{a^* T_s^*} L^*)^{-2} ds. \tag{7.29}$$

We note that

$$\hat{T}_t^* = \sigma^2 \int_0^t (S_0 - L^* + \tilde{W}^*((\hat{T}_t^*) + e^{a^* s} L^*)^2 ds, \tag{7.30}$$

where a^* and L^* are defined in (7.27).

7.4.2 Some Properties of the Process $\tilde{W}^*(\hat{T}_t^*)$

Using the same argument as in Section 7.4, we obtain the following properties of the process $\tilde{W}^*((\hat{T}_t^*)$ in (7.25). This is a zero-mean P^*-martingale and

$$E^* \tilde{W}^*((\hat{T}_t^*) = 0,$$
$$E^* [\tilde{W}^*((\hat{T}_t^*)]^2 = \sigma^2 [(S_0 - L^*)^2 \frac{e^{\sigma^2 t} - 1}{\sigma^2} + \frac{2L^*(S_0 - L^*)(e^{a^* t} - e^{\sigma^2 t})}{a^* - \sigma^2}$$
$$+ \frac{(L^*)^2 (e^{2a^* t} - e^{\sigma^2 t})}{2a^* - \sigma^2}], \tag{7.31}$$

where E^* is the expectation with respect to the probability P^* and a^* and L^* and $(\phi_t^*)^{-1}$ are defined in (7.27) and (7.30), respectively.

7.4.3 The Explicit Expression for the Process $\tilde{W}^*(\hat{T}_t)$

It turns out that we can find the explicit expression for the process $\tilde{W}^*(\hat{T}_t)$.
From the expression

$$V_t = S_0 - L + \tilde{W}^*(\hat{T}_t),$$

we have the following relationship between $W(t)$ and $\tilde{W}(\hat{T}_t)$:

$$d\tilde{W}^*(\hat{T}_t) = \sigma \int_0^t [S(0) - L + Le^{at} + \tilde{W}^*(\hat{T}_t)] dW^*(t).$$

It is a linear SDE with respect to $\tilde{W}^*(\hat{T}_t)$ and we can solve it explicitly. The solution has the following form:

$$\tilde{W}^*(\hat{T}_t) = S(0)(e^{\sigma W^*(t) - \frac{\sigma^2 t}{2}} - 1) + L(1 - e^{at})$$
$$+ aLe^{\sigma W^*(t) - \frac{\sigma^2 t}{2}} \int_0^t e^{as} e^{-\sigma W^*(s) + \frac{\sigma^2 s}{2}} ds. \tag{7.32}$$

From (7.32) it is easy to see that $\tilde{W}^*(\hat{T}_t)$ can be presented in the form of a linear combination of two zero-mean P^*-martingales $m_1^*(t)$ and $m_2^*(t)$:

$$\tilde{W}^*(\hat{T}_t) = m_1^*(t) + L^* m_2^*(t),$$

where

$$m_1^*(t) := S(0)(e^{\sigma W^*(t) - \frac{\sigma^2 t}{2}} - 1)$$

and

$$m_2^*(t) = (1 - e^{a^* t}) + a^* e^{\sigma W^*(t) - \frac{\sigma^2 t}{2}} \int_0^t e^{a^* s} e^{-\sigma W^*(s) + \frac{\sigma^2 s}{2}} ds.$$

Indeed, process $\tilde{W}^*(\hat{T}_t)$ is a martingale (see Section 5.2); also it is well known that process $e^{\sigma W^*(t) - \frac{\sigma^2 t}{2}}$, and, hence, process $m_1^*(t)$ is a martingale. Then the process $m_2^*(t)$, as the difference between two martingales, is also martingale. In this way, we have

$$E_{P^*} m_1^*(t) = 0,$$

since

$$E_{P^*} e^{\sigma W^*(t) - \frac{\sigma^2 t}{2}} = 1.$$

As for $m_2(t)$, we have

$$E_{P^*} m_2(t) = 0,$$

since from Itô's formula, we have

$$\begin{aligned}
d\ &(a^* e^{\sigma W^*(t) - \frac{\sigma^2 t}{2}} \int_0^t e^{a^* s} e^{-\sigma W^*(s) + \frac{\sigma^2 s}{2}} ds) \\
&= a^* \sigma e^{\sigma W^*(t) - \frac{\sigma^2 t}{2}} \int_0^t e^{a^* s} e^{-\sigma W^*(s) + \frac{\sigma^2 s}{2}} ds dW^*(t) \\
&\quad + a^* e^{\sigma W^*(t) - \frac{\sigma^2 t}{2}} e^{a^* t} e^{-\sigma W^*(t) + \frac{\sigma^2 t}{2}} dt \\
&= a^* \sigma e^{\sigma W^*(t) - \frac{\sigma^2 t}{2}} \int_0^t e^{a^* s} e^{-\sigma W^*(s) + \frac{\sigma^2 s}{2}} ds dW^*(t) \\
&\quad + a^* e^{a^* t} dt,
\end{aligned}$$

and, hence,

$$E_{P^*} a^* e^{\sigma W^*(t) - \frac{\sigma^2 t}{2}} \int_0^t e^{a^* s} e^{-\sigma W^*(s) + \frac{\sigma^2 s}{2}} ds = e^{a^* t} - 1.$$

It is interesting to see that in the last expression, the first moment for

$$\eta^*(t) := a^* e^{\sigma W^*(t) - \frac{\sigma^2 t}{2}} \int_0^t e^{a^* s} e^{-\sigma W^*(s) + \frac{\sigma^2 s}{2}} ds,$$

does not depend on σ.

This is true not only for the first moment but for all the moments of the process $\eta^*(t) = a^* e^{\sigma W^*(t) - \frac{\sigma^2 t}{2}} \int_0^t e^{a^* s} e^{-\sigma W^*(s) + \frac{\sigma^2 s}{2}} ds.$

Indeed, using Itô's formula for $(\eta^*(t))^n$, we obtain

$$d(\eta^*(t))^n = n(a^*)^n \sigma e^{n\sigma W^*(t) - \frac{n\sigma_t^2}{2}} (\int_0^t e^{a^*s} e^{-\sigma W^*(s) + \frac{\sigma_s^2}{2}} ds)^n dW^*(t)$$
$$+ a^* n(\eta_2(t))^{n-1} e^{a^*t} dt,$$

and

$$dE(\eta^*(t))^n = na^* e^{a^*t} E(\eta^*(t))^{n-1} dt, \quad n \geq 1.$$

This is a recursive equation with the initial function $(n = 1)$ $E\eta^*(t) = e^{a^*t} - 1$. After calculations we obtain the following formula for $E(\eta^*(t))^n$:

$$E(\eta^*(t))^n = (e^{a^*t} - 1)^n.$$

7.4.4 Some Properties of the Mean-Reverting Risk-Neutral Asset S_t

Using the same argument as in Section 7.5, we obtain the following properties of the mean-reverting risk-neutral asset S_t in (7.18):

$$E^* S_t = e^{-a^*t}[S_0 - L^*] + L^*$$
$$Var^*(S_t) := E^* S_t^2 - (E^* S_t)^2$$
$$= \sigma^2 e^{-2a^*t}[(S_0 - L^*)^2 \frac{e^{\sigma^2 t} - 1}{\sigma^2} + \frac{2L^*(S_0 - L^*)(e^{a^*t} - e^{\sigma^2 t})}{a^* - \sigma^2} \qquad (7.33)$$
$$+ \frac{(L^*)^2(e^{2a^*t} - e^{\sigma^2 t})}{2a^* - \sigma^2}],$$

where E^* is the expectation with respect to the probability P^* and $a^* and L^*$ and $(\phi_t^*)^{-1}$ are defined in (7.27) and (7.30), respectively.

From the expression for $\tilde{W}^*(\phi_t^{-1})$ (see (7.32)) and for $S(t)$ in (7.28) (see also (7.29)–(7.30)), we can find the explicit expression for $S(t)$ through $W^*(t)$:

$$S(t) = e^{-a^*t}[S_0 - L^* + \tilde{W}^*(\hat{T}_t)] + L^*$$
$$= e^{-a^*t}[S_0 - L^* + m_1^*(t) + L^* m_2^*(t)] + L^* \qquad (7.34)$$
$$= S(0)e^{-at} e^{\sigma W^*(t) - \frac{\sigma_t^2}{2}} + aLe^{-at} e^{\sigma W^*(t) - \frac{\sigma_t^2}{2}} \int_0^t e^{as} e^{-\sigma W^*(s) + \frac{\sigma_s^2}{2}} ds,$$

where $m_1^*(t)$ and $m_2^*(t)$ are defined as in Section 5.3.

7.4.5 The Explicit Option Pricing Formula for the European Call Option for the Mean-Reverting Asset Model Under Risk-Neutral Measure

Proceeding with the same calculations (7.15)–(7.22) as in Section 7.3, where in place of a and L we take a^* and L^* in (7.27), we obtain the following theorem.

Theorem 5.1. *The explicit option pricing formula for the European call option under risk-neutral measure has the following form:*

$$C_T^* = e^{-(r+a^*)T}S(0)\Phi(y_+) - e^{-rT}K\Phi(y_-)$$
$$+ L^*e^{-(r+a^*)T}[(e^{a^*T} - 1) - \int_0^{y_0} zF_T^*(dz)], \tag{7.35}$$

where y_0 is the solution of the following equation:

$$y_0 = \frac{\ln(\frac{K}{S(0)}) + (\frac{\sigma^2}{2} + a^*)T}{\sigma\sqrt{T}}$$

$$- \frac{\ln(1 + \frac{a^*L^*}{S(0)}\int_0^T e^{a^*s}e^{-\sigma y_0\sqrt{s} + \frac{\sigma^2 s}{2}}ds)}{\sigma\sqrt{T}}, \tag{7.36}$$

$$y_+ := \sigma\sqrt{T} - y_0 \quad and \quad y_- := -y_0, \tag{7.37}$$

$$a^* := a + \lambda\sigma, \quad L^* := \frac{aL}{a + \lambda\sigma},$$

and $F_T^*(dz)$ is the probability distribution as in (7.21), where instead of parameter a, we have to take $a^* = a + \lambda\sigma$.

Remark. From (7.35) we can find that the European call option price C_T^* for the mean-reverting asset under risk-neutral measure lies between the following boundaries:

$$e^{-(r+a^*)T}S(0)\Phi(y_+) - e^{-rT}K\Phi(y_-) \leq C_T$$
$$\leq e^{-(r+a^*)T}S(0)\Phi(y_+) - e^{-rT}K\Phi(y_-)$$
$$+ L^*e^{-(r+a^*)T}[e^{a^*T} - 1], \tag{7.38}$$

where $y_0, y_-,$ and y_+ are defined in (7.36)–(7.37).

7.4.6 The Black-Scholes Formula Follows $L^* = 0$ and $a^* = -r$!

If $L^* = 0$ and $a^* = -r$, then, from (7.35), we obtain

$$C_T = S(0)\Phi(y_+) - e^{-rT}K\Phi(y_-), \tag{7.39}$$

where

$$y_+ := \sigma\sqrt{T} - y_0 \quad and \quad y_- := -y_0, \tag{7.40}$$

and y_0 is the solution of the following equation (see (7.36))

$$S(0)e^{-rT}e^{\sigma y_0\sqrt{T} - \frac{\sigma^2 T}{2}} = K$$

or

$$y_0 = \frac{\ln(\frac{K}{S(0)}) + (\frac{\sigma^2}{2} - r)T}{\sigma\sqrt{T}}. \tag{7.41}$$

But (7.39)–(7.41) is the exact well-known Black-Scholes result!

7.5 A Numerical Example: The AECO Natural GAS Index (1 May 1998–30 April 1999)

We shall calculate the value of European call option on the price of a daily natural gas contract. To apply our formula for calculating this value, we need to calibrate the parameters a, L, σ, and λ. These parameters may be obtained from futures prices for the AECO Natural Gas Index for the period 1 May 1998 to 30 April 1999 (see Bos et al. 2002, p.340). The parameters pertaining to the option are presented in Table 7.1.

Table 7.1 Price and option process parameters

Price and option process parameters						
T	a	σ	L	λ	r	K
6 months	4.6488	1.5116	2.7264	0.1885	0.05	3

From this table we can calculate the values for a^* and L^*.

$$a^* = a + \lambda\sigma = 4.9337,$$

and

$$L^* = \frac{aL}{a + \lambda\sigma} = 2.5690.$$

For the value of S_0, we can take $S_0 \in [1,6]$.

Figure 7.1 (see Appendix) depicts the dependence of the mean value ES_t on the maturity T for AECO Natural Gas Index (1 May 1998 to 30 April 1999).

Figure 7.2 (see Appendix) depicts the dependence of the mean value ES_t on the initial value of stock S_0 and maturity T for AECO Natural Gas Index (1 May 1998 to 30 April 1999).

Figure 7.3 (see Appendix) depicts the dependence of the variance of S_t on the initial value of stock S_0 and maturity T for AECO Natural Gas Index (1 May 1998 to 30 April 1999).

Figure 7.4 (see Appendix) depicts the dependence of the volatility of S_t on the initial value of stock S_0 and maturity T for AECO Natural Gas Index (1 May 1998 to 30 April 1999).

Figure 7.5 (see Appendix) depicts the dependence of the European call option price for MRRNAM on the maturity (months) for AECO Natural Gas Index (1 May 1998 to 30 April 1999) with $S(0) = 1$ and $K = 3$.

Appendix: Figures

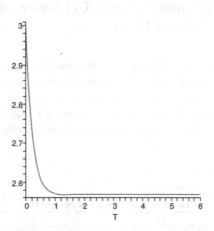

Fig. 7.1 Dependence of ES_t on T (AECO Natural Gas Index (1 May 1998–30 April 1999))

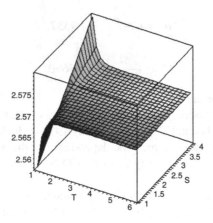

Fig. 7.2 Dependence of ES_t on S_0 and T (AECO Natural Gas Index (1 May 1998–30 April 1999))

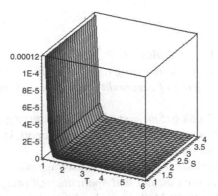

Fig. 7.3 Dependence of variance of S_t on S_0 and T (AECO Natural Gas Index (1 May 1998–30 April 1999))

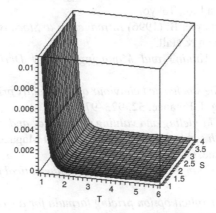

Fig. 7.4 Dependence of volatility of S_t on S_0 and T (AECO Natural Gas Index (1 May 1998–30 April 1999))

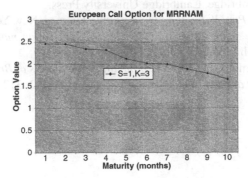

Fig. 7.5 Dependence of European call option price on maturity (months) ($S(0) = 1$ and $K = 3$) (AECO Natural Gas Index (1 May 1998–30 April 1999))

References

Bos, L. P., Ware, A. F. and Pavlov, B. S. (2002) *On a semi-spectral method for pricing an option on a mean-reverting asset*, Quantitative Finance, **2**, 337–345.

Black, F. (1976) *The pricing of commodity contracts*, J. Financial Economics, **3**, 167–179.

Chen, Z. and Forsyth, P. (2006) *Stochastic models of natural gas prices and applications to natural gas storage valuation*, Technical Report, University of Waterloo, Waterloo, Canada, November 24, 32p.
   (http://www.cs.uwaterloo.ca/ paforsyt/regimestorage.pdf).

Elliott, R. (1982) *Stochastic Calculus and Applications*, Springer-Verlag, New York.

Elliott, R. and Kopp, E. (1999) *Mathematics of Financial Markets*, Springer-Verlag, New York.

Ikeda, N. and Watanabe, S. (1981) *Stochastic Differential Equations and Diffusion Processes*, Kodansha Ltd., Tokyo.

Lamperton, D. and Lapeyre, B. (1996) *Introduction to Stochastic Calculus Applied to Finance*, Chapmann & Hall.

Pilipovic, D. (1997) *Valuing and Managing Energy Derivatives*, New York, McGraw-Hill.

Schwartz, E. (1997) *The stochastic behaviour of commodity prices: implications for pricing and hedging*, J. Finance, **52**, 923–973.

Swishchuk, A. (2004) *Modeling and valuing of variance and volatility swaps for financial markets with stochastic volatilities*, Wilmott Magazine, Technical Article No2, September Issue, 64–72.

Swishchuk, A. (2007) *Change of time method in mathematical finance*, CAMQ, Vol. 15, No. 3, 299–336.

Swishchuk A. (2008) *Explicit option pricing formula for a mean-reverting asset in energy market*. Journal of Numerical and Applied Mathematics. 1(96): 23.

Wilmott, P., Howison, S. and Dewynne, J. (1995) *The Mathematics of Financial Derivatives*, Cambridge, Cambridge University Press.

Wilmott, P. (2000) *Paul Wilmott on Quantitative Finance*, New York, Wiley.

Yor, M. (1992) *On some exponential functions of Brownian motion*, Advances in Applied Probability, Vol. 24, No. 3, 509–531.

Yor, M. and Matsumoto, H. (2005) *Exponential Functionals of Brownian motion, I: Probability laws at fixed time*, Probability Surveys, Vol. 2, 312–347.

Chapter 8
CTM and Multifactor Lévy Models for Pricing Financial and Energy Derivatives

"The energy of the mind is the essence of life". —Aristotle.

Abstract In this chapter, the CTM is applied to price financial and energy derivatives for one-factor and multifactor α-stable Lévy-based models. These models include, in particular, as one-factor models, the Lévy-based geometric motion model and the Ornstein and Uhlenbeck (1930), the Vasicek (1977), the Cox et al. (1985), the continuous-time GARCH, the Ho and Lee (1986), the Hull and White (1990), and the Heath et al. (1992) models and, as multifactor models, various combinations of the previous models. For example, we introduce new multifactor models such as the Lévy-based Heston model, the Lévy-based SABR/LIBOR market models, and Lévy-based Schwartz-Smith and Schwartz models. Using the change of time method for SDEs driven by α-stable Lévy processes, we present the solutions of these equations in simple and compact forms. We then apply this method to price many financial and energy derivatives such as variance swaps, options, forward, and futures contracts.

8.1 Introduction

In this section, we will first review the change of time method (CTM) for Lévy-based models and give an overview of the multifactor Gaussian, LIBOR and SABR models, swaps, and energy derivatives.

© The Author 2016
A. Swishchuk, *Change of Time Methods in Quantitative Finance*,
SpringerBriefs in Mathematics, DOI 10.1007/978-3-319-32408-1_8

8.1.1 A Change of Time Method (CTM) for Lévy-Based Models: Short Literature History

Rosinski and Woyczinski (1986) considered time changes for integrals over a stable Lévy process. Kallenberg (1992) considered time change representations for stable integrals.

Lévy processes can also be used as a time change for other Lévy processes (subordinators). Madan and Seneta (1990) introduced the variance gamma (VG) process (Brownian motion with drift time changed by a gamma process). Geman et al. (2001) considered time changes ("business times") for Lévy processes. Carr et al. (2003) used a change of time to introduce stochastic volatility into a Lévy model to achieve a leverage effect and a long-term skew. Kallsen and Shiryaev (2001) showed that the Rosiński-Woyczyński-Kallenberg result cannot be extended to any other Lévy processes other than the symmetric α-stable processes. Swishchuk (2004, 2007) applied a change of time method for option and swap pricing for Gaussian models.

The book *Change of Time and Change of Measure* by Barndorff-Nielsen and Shiryaev (2010) states the main ideas and results of the stochastic theory of "change of time and change of measure" in the semimartingale setting.

8.1.2 Stochastic Differential Equations (SDEs) Driven by Lévy Processes

Girsanov (1960) used the change of time method to construct a weak solution to a specific SDE driven by Brownian motion.

The existence and uniqueness of the solutions for SDEs driven by Lévy processes have been studied in Applebaum (2003). The existence and uniqueness of the solutions for SDEs driven by general semimartingale with jumps have been studied in Protter (2005) and Jacod (1979).

Janicki et al. (1996) proved that there exists a unique solution of the SDE for continuous drift b_s, the diffusion coefficient σ, and α-stable Lévy process $S_\alpha((t-s)^{1/\alpha}, \beta, \delta), \beta \in [-1, +1]$.

Zanzotto (1997) had also considered solutions of one-dimensional SDEs driven by stable Lévy motion.

Cartea and Howison (2006) considered option pricing with Lévy-stable processes generated by Lévy-stable integrated variance.

8.1.3 LIBOR Market and SABR Models: Short Literature Review

The basic log-normal forward LIBOR (also known as LIBOR Market, or BGM) model has proved to be an essential tool for pricing and risk-managing interest rate derivatives. First introduced in Brace et al. (1997), and Jamshidian (1997), forward LIBOR models are in the mainstream of interest rate modelling. For general information regarding the model, textbooks such as Brigo and Mercurio (2001) and Rebonato (2002) provide a good starting point. Various extensions of forward LIBOR models that attempt to incorporate volatility smiles of interest rates have been proposed. Local volatility type extensions were pioneered in Andersen and Andersen (2000). A stochastic volatility extension is proposed in Andersen and Brotherton-Ratcliffe (2001) and further extended in Andersen and Andersen (2002). A different approach to stochastic volatility forward LIBOR models is described in Rebonato (2002). Jump-diffusion forward LIBOR models are treated in Glasserman and Merener (2003) and Glasserman and Kou (1999). In Sin (2002), a stochastic volatility/jump-diffusion forward LIBOR model is advocated.

The SABR and the LIBOR market models (LMM) have become industry standards for pricing plain-vanilla and complex interest rate products, respectively. For a description of the SABR model, see, for example, Hagan et al. (2002). Several stochastic volatility extensions of the LMM exist that do provide a consistent dynamic description of the evolution of the forward rates (e.g. see Andersen and Andersen 2000; Joshi and Rebonato 2003; Rebonato and Joshi 2002; Rebonato and Kainth 2004), but these extensions are not equivalent to the SABR model. Piterbarg (2003, 2005) presents an approach based on displaced diffusion that is similar in spirit to a dynamical extension of the SABR model. Henry-Labordeè (2007) obtains some interesting exact results, but his attempt to unify the BGM and SABR models using an application of hyperbolic geometry is very complex and the computational issues are daunting. Rebonato (2007) proposed an extension of the LMM that recovers the SABR caplet prices almost exactly for all strikes and maturities. Many smiles and skews are usually managed by using local volatility models by Dupire (1994).

8.1.4 Energy Derivatives' Overview

Black's model (1976) and Schwartz's model (1997) have become a standard approach to the problem of pricing options on commodities. These models have the advantage of mathematical convenience, in that they give rise to closed-form solutions for some types of options (see Wilmott 2000).

A drawback of single-factor mean-reverting models lies in the case of option pricing: the fact the long-term rate is fixed results in a model-implied volatility term structure that has the volatilities going to zero as the expiration time increases.

Using single-factor non-mean-reverting models also has a drawback: it will impact valuation and hedging. The differences between the distributions are particularly obvious when pricing out-of-the-money options, where the tails of the distribution play a very important role. Thus, if a log-normal model, for example, is used to price a far out-of-the-money option, the price can be very different from a mean-reverting model's price (see Pilipovic (1998)). A popular model used for modelling energy and agricultural commodities and introduced by Schwartz (1990) aims at resembling the geometric Brownian motion while introducing mean reversion to a long-term value in the drift term (see Schwartz 1997). This mean-reverting model is a one-factor version of the two-factor model made popular in the context of energy modelling by Pilipovic (1997).

Villaplana (2004) proposed the introduction of two sources of risk X and Y representing, respectively, short-term and long-term shocks, and describes the spot price S_t. Geman and Roncoroni (2005) (see also Geman (2005)) introduced a jump-reversion model for electricity prices. The two-factor model for oil-contingent claim pricing was proposed by Gibson and Schwartz (1990). Eydeland and Geman (1998) proposed extending the Heston (1993) stochastic volatility model to gas or electricity prices by introducing mean reversion in the spot price and proposing two-factor model. Geman (2000) introduced three-factor model for commodity prices taking into account stochastic equilibrium level and stochastic volatility. Björk and Landen (2002) investigated the term structure of forward and futures prices for models where the price processes are allowed to be driven by a general market point process as well as by a multidimensional Wiener process. Benth et al. (2008) applied independent increments processes (see Skorokhod 1964; Lévy 1965) to model and price electricity, gas, and temperature derivatives (forwards, futures, swaps, options). Swishchuk (2008) considers a risky asset in energy markets following mean-reverting stochastic process. An explicit expression for a European option price based on this asset, using a change of time method, is derived. A numerical example for the AECO Natural Gas Index (1 May 1998–30 April 1999) is presented.

8.2 α-Stable Lévy Processes and Their Properties

8.2.1 Lévy Processes

DEFINITION 2. By Lévy process we mean a stochastically continuous process with stationary and independent increments (see Sato 2005; Applebaum 2003; Schoutens 2003).

Examples of Lévy Processes $L(t)$ include a linear deterministic function $L(t) = \gamma t$; Brownian motion with drift; Poisson process, compound Poisson process; jump-diffusion process; and variance gamma (VG), inverse Gaussian (IG), normal inverse Gaussian (NIG), generalized hyperbolic, and α-stable processes (see Sato (2005)).

8.2.2 Lévy-Khintchine Formula and Lévy-Itô Decomposition for Lévy Processes $L(t)$

The characteristic function of the Lévy process follows the following formula (so-called Lévy-Khintchine formula):

$$E(e^{i(u,L(t))}) = \exp\{t[i(u,\gamma) - \tfrac{1}{2}(u,Au) + \int_{R^d-\{0\}}[e^{i(u,y)} - 1 - i(u,y)\mathbf{1}_{B_1(0)}]\nu(dy)]\}$$

where (γ, A, ν) is the Lévy-Khintchine triplet.

If L is a Lévy process, then there exists $\gamma \in R^d$, a Brownian motion B_A with co-variance matrix A and an independent Poisson random measure N on $R^+ \times (R^d - \{0\})$ such that for each $t \geq 0$, $L(t)$ has the following decomposition (Lévy-Itô decomposition):

$$L(t) = \gamma t + B_A(t) + \int_{|x|<1} x\tilde{N}(t, dx) + \int_{|x|\geq 1} xN(t, dx),$$

where N is a Poisson counting measure and $\tilde{N}$ is a compensated Poisson measure (see Applebaum 2003).

Remark (Lévy Processes in Finance). The most commonly used Lévy processes in finance include Brownian motion with drift (the only continuous Lévy process), the Merton process=Brownian motion+drift+Gaussian jumps, the Kou process=Brownian motion+drift+exponential jumps and variance gamma (VG), inverse Gaussian (IG), normal inverse Gaussian (NIG), generalized hyperbolic (GH) and α-stable Lévy.

8.2.3 α-Stable Distributions and Lévy Processes

In this section, we introduce α-stable distributions and Lévy processes and describe their properties.

8.2.3.1 Symmetric α-Stable ($S\alpha S$) Distribution

The characteristic function of the $S\alpha S$ distribution is defined as follows:

$$\phi(u) = e^{(i\delta u - \sigma|u|^\alpha)},$$

where α is the *characteristic exponent* ($0 < \alpha \leq 2$), $\delta \in (-\infty, +\infty)$ is the *location* parameter, and $\sigma > 0$ is the *dispersion*.

For values of $\alpha \in (1,2]$ the location parameter δ corresponds to the mean of the α-stable distribution, while for $0 < \alpha \leq 1$, δ corresponds to its median.

The dispersion parameter σ corresponds to the spread of the distribution around its location parameter δ. The characteristic exponent α determines the shape of the distribution.

A stable distribution is called *standard* if $\delta = 0$ and $\sigma = 1$. If a random variable L is stable with parameters $\alpha, \delta, \sigma,$, then $(L - \delta)/\sigma^{1/\alpha}$ is standard with characteristic exponent α. By letting α take the values 1/2, 1, and 2, we get three important special cases: the Lévy ($\alpha = 1/2$), Cauchy ($\alpha = 1$), and the Gaussian ($\alpha = 2$) distributions:

$$f_{1/2}(\gamma, \delta; x) = (\tfrac{t}{2\sqrt{\pi}})x^{-3/2}e^{-t^2/(4x)}$$
$$f_1(\gamma, \delta; x) = \tfrac{1}{\pi}\tfrac{\gamma}{\gamma^2 + (x-\delta)^2},$$
$$f_2(\gamma, \delta; x) = \tfrac{1}{\sqrt{4\pi\gamma}}\exp[-\tfrac{(x-\delta)^2}{4\gamma}].$$

Unfortunately, no closed-form expression exists for general α-stable distributions other than the Lévy, the Cauchy, and the Gaussian. However, power series expansions can be derived for the density $f_\alpha(\delta, \sigma; x)$. Its tails (algebraic tails) decay at a lower rate than the Gaussian density tails (exponential tails).

The smaller the characteristic exponent α is, the heavier the tails of the α-stable density.

This implies that random variables following α-stable distribution with small characteristic exponent are *highly impulsive*, and it is this heavy-tail characteristic that makes this density appropriate for modelling noise which is impulsive in nature, for example, electricity prices or volatility.

Only moments of order less than α exist for the non-Gaussian family of α-stable distribution. The fractional lower-order moments with zero location parameter and dispersion σ are given by

$$E|X|^p = D(p, \alpha)\sigma^{p/\alpha}, \quad for \quad 0 < p < \alpha,$$
$$D(p, \alpha) = \tfrac{2^p \Gamma(\frac{p+1}{2})\Gamma(1-\frac{p}{\alpha})}{\alpha\sqrt{\pi}\Gamma(1-\frac{p}{2})},$$

where $\Gamma(\cdot)$ is the gamma function (Sato (1999)).

Since the $S\alpha S$ r.v. has "infinite variance", the covariation of two jointly $S\alpha S$ real r.v. with dispersions γ_x and γ_y defined by

$$[X, Y]_\alpha = \frac{E[X|Y|^{p-2}Y]}{E[|Y|^p]}\gamma_y$$

has often been used instead of the covariance (and correlation), where $\gamma_y = [Y, Y]_\alpha$ is the dispersion of r.v. Y.

8.2.3.2 α-Stable Lévy Processes

DEFINITION 3. Let $\alpha \in (0, 2]$. An α-STABLE LÉVY PROCESS L such that L_1 (or equivalently any L_t) has a strictly α-stable distribution (i.e. $L_1 \equiv S_\alpha(\sigma, \beta, \delta)$) for

some $\alpha \in (0,2] \setminus \{1\}$, $\sigma \in R_+$, $\beta \in [-1,1]$, $\delta = 0$ or $\alpha = 1$, $\sigma \in R_+$, $\beta = 0$, $\delta \in R$).
We call L a SYMMETRIC α-STABLE LÉVY PROCESS if the distribution of L_1 is even
symmetric α-stable (i.e. $L_1 \equiv S_\alpha(\sigma,0,0)$ for some $\alpha \in (0,2]$, $\sigma \in R_+$.) A process
L is called $(T_t)_{t \in R_+}$ adapted if L is constant on $[T_{t-}, T_t]$ for any $t \in R_+$ (see Sato
(1999)).

8.2.3.3 Properties of α-Stable Lévy Processes

The α-stable Lévy processes are the only self-similar Lévy processes such that
$L(at) \overset{Law}{=} a^{1/\alpha} L(t), a \geq 0$. They are either Brownian motion or pure jump. They have
characteristic exponent and Lévy-Khintchine triplet known in closed form. They
also have only four parameters, but infinite variance (except for Brownian motion).
The α-stable Lévy processes are semimartingales (in this way, $\int_0^t f_s dL_s$ can be defined), and α-stable Lévy processes are pure discontinuous Markov processes with
generator

$$Af(x) = \int_{R-\{0\}} [f(x+y) - f(x) - yf'(y)\mathbf{1}_{|y|<1}(y)] \frac{K_\alpha}{|y|^{1+\alpha}} dy.$$

$E|L(t)|^p$ is finite or infinite accordingly as $0 < p < \alpha$ or $p > \alpha$, respectively. In
particular, for an α-stable process, $EL(t) = \delta t$ ($1 < \alpha < 2$) (Sato (1999)).

8.3 Stochastic Differential Equations Driven by α-Stable Lévy Processes

Consider the following SDE driven by an α-stable Lévy process $L(t)$:

$$dZ_t = b(t, Z_{t-})dt + \sigma(t, Z_{t-})dL(t). \tag{8.1}$$

Janicki et al. (1996) proved that this equation has a weak solution for continuous
coefficients a and b.

We consider below one-factor and multifactor models described by SDEs driven
by α-stable Lévy process $L(t)$.

8.3.1 One-Factor α-Stable Lévy Models

$L(t)$ below is a symmetric α-stable Lévy process. We define below various processes via SDE driven by α-stable Lévy process:

1. The geometric α-stable Lévy motion: $dS(t) = \mu S(t-)dt + \sigma S(t-)dL(t)$.
2. The Ornstein-Uhlenbeck process driven by α-stable Lévy motion: $dS(t) = -\mu S(t-)dt + \sigma dL(t)$.

3. The Vasićek process driven by α-stable Lévy motion: $dS(t) = \mu(b - S(t-))dt + \sigma dL(t)$.
4. The continuous-time GARCH process driven by α-stable Lévy motion: $dS(t) = \mu(b - S(t-))dt + \sigma S(t-)dL(t)$.
5. The Cox-Ingersoll-Ross process driven by α-stable Lévy motion: $dS(t) = k(\theta - S(t-))dt + \gamma\sqrt{S(t-)}dL(t)$.
6. The Ho and Lee process driven by α-stable Lévy motion: $dS(t) = \theta(t-)dt + \sigma dL(t)$.
7. The Hull and White process driven by α-stable Lévy motion: $dS(t) = (a(t-) - b(t-)S(t-))dt + \sigma(t)dL(t)$
8. The Heath, Jarrow, and Morton process driven by α-stable Lévy motion: Define the forward interest rate $f(t, s)$, for $t \leq s$, that represents the instantaneous interest rate at time s as "anticipated" by the market at time t. The process $f(t, u)_{0 \leq t \leq u}$ satisfies an equation

$$f(t, u) = f(0, u) + \int_0^t a(v, u)dv + \int_0^t b(f(v, u))dL(v),$$

where the processes a and b are continuous.

We note that Eberlein and Raible (1999) considered Lévy-based term structure models.

8.3.2 Multifactor α-Stable Lévy Models

Multifactor models driven by α-stable Lévy motions can be obtained using various combinations of the above-mentioned processes. We give one example of a two-factor continuous-time GARCH model driven by α-stable Lévy motions:

$$dS(t) = r(t-)S(t-))dt + \sigma S(t-)dL^1(t)$$
$$dr(t) = a(m - r(t-))dt + \sigma_2 r(t-)dL^2(t),$$

where L^1 and L^2 may be correlated, $m \in R, \sigma_i, a > 0, i = 1, 2$.

Also, we can consider various combinations of models, presented above, i.e. mixed models containing Brownian and Lévy motions. For example,

$$dS(t) = \mu(b(t-) - S(t-))dt + \sigma S(t-)dL(t)$$
$$db(t) = \xi b(t)dt + \eta b(t)dW(t),$$

where the Brownian motion $W(t)$ and Lévy process $L(t)$ may be correlated.

8.4 Change of Time Method (CTM) for SDEs Driven by Lévy Processes

We denote by $L_{a.s.}^{\alpha}$ the family of all real measurable $\mathscr{F}_t$-adapted processes a on $\Omega \times [0, +\infty)$ such that for every $T > 0$, $\int_0^T |a(t, \omega)|^{\alpha} dt < +\infty$ a.s. We consider the following SDE driven by a Lévy motion:

$$dX(t) = a(t, X(t-))dL(t),$$

where $L(t)$ is an α-stable Lévy process.

Theorem (Rosinski and Woyczinski (1986), Theorem 3.1., p. 277). Let $a \in L_{a.s.}^{\alpha}$ be such that $T(u) := \int_0^u |a|^{\alpha} dt \to +\infty$ a.s. as $u \to +\infty$. If $\hat{T}(t) := \inf\{u : T(u) > t\}$ and $\hat{\mathscr{F}}_t = \mathscr{F}_{\hat{T}(t)}$, then the time-changed stochastic integral $\hat{L}(t) = \int_0^{\hat{T}(t)} adL(t)$ is an $\hat{\mathscr{F}}_t - \alpha$-stable Lévy process, where $L(t)$ is $\mathscr{F}_t$-adapted and $\mathscr{F}_t$-α-stable Lévy process. Consequently, a.s. for each $t > 0$ $\int_0^t adL = \hat{L}(T(t))$, i.e. the stochastic integral with respect to a α-stable Lévy process is nothing but another α-stable Lévy process with randomly changed timescale.

8.4.1 Solutions of One-Factor Lévy Models Using the CTM

Below we give the solutions to the one-factor Lévy models described by SDEs driven by α-stable Lévy process introduced in Section 8.2.3.

Proposition 1. Let $L(t)$ be a symmetric α-stable Lévy process, and $\hat{L}$ is a $(\hat{T}_t)_{t \in R_+}$-adapted symmetric α-stable Levy process on $(\Omega, \mathscr{F}, (\hat{\mathscr{F}}_t)_{t \in R_+}, P))$. Then, we have the following solutions for the above-mentioned one-factor Lévy models 1–8 (Section 8.3.1):

1. The geometric α-stable Lévy motion: $dS(t) = \mu S(t-)dt + \sigma S(t-)dL(t)$. Solution $S(t) = e^{\mu t}[S(0) + \hat{L}(\hat{T}_t)]$, where $\hat{T}_t = \sigma^{\alpha} \int_0^t [S(0) + \hat{L}(\hat{T}_s)]^{\alpha} ds$.
2. The Ornstein-Uhlenbeck process driven by α-stable Lévy motion: $dS(t) = -\mu S(t-)dt + \sigma dL(t)$. Solution $S(t) = e^{-\mu t}[S(0) + \hat{L}(\hat{T}_t)]$, where $\hat{T}_t = \sigma^{\alpha} \int_0^t (e^{\mu s}[S(0) + \hat{L}(\hat{T}_s)])^{\alpha} ds$.
3. The Vasiček process driven by α-stable Lévy motion: $dS(t) = \mu(b - S(t-))dt + \sigma dL(t)$. Solution $S(t) = e^{-\mu t}[S(0) - b + \hat{L}(\hat{T}_t)]$, where $\hat{T}_t = \sigma^{\alpha} \int_0^t (e^{\mu s}[S(0) - b + \hat{L}(\hat{T}_s)] + b)^{\alpha} ds$.
4. The continuous-time GARCH process driven by α-stable Lévy process: $dS(t) = \mu(b - S(t-))dt + \sigma S(t-)dL(t)$. Solution $S(t) = e^{-\mu t}(S(0) - b + \hat{L}(\hat{T}_t)) + b$, where $\hat{T}_t = \sigma^{\alpha} \int_0^t [S(0) - b + \hat{L}(\hat{T}_s) + e^{\mu s}b]^{\alpha} ds$.
5. The Cox-Ingersoll-Ross process driven by α-stable Lévy motion: $dS(t) = k(\theta^2 - S(t-))dt + \gamma\sqrt{S(t-)}dL(t)$. Solution $S^2(t) = e^{-kt}[S_0^2 - \theta^2 + \hat{L}(\hat{T}_t)] + \theta^2$, where $\hat{T}_t = \gamma^{\alpha} \int_0^t [e^{k\hat{T}_s}(S_0^2 - \theta^2 + \hat{L}(\hat{T}_s)) + \theta^2 e^{2k\hat{T}_s}]^{\alpha/2} ds$.

6. The Ho and Lee process driven by α-stable Lévy motion: $dS(t) = \theta(t-)dt + \sigma dL(t)$. Solution $S(t) = S(0) + \hat{L}(\sigma^\alpha t) + \int_0^t \theta(s)ds$.

7. The Hull and White process driven by α-stable Lévy motion: $dS(t) = (a(t-) - b(t-)S(t-))dt + \sigma(t-)dL(t)$. Solution $S(t) = \exp[-\int_0^t b(s)ds][S(0) - \frac{a(s)}{b(s)} + \hat{L}(\hat{T}_t)]$,
   where $\hat{T}_t = \int_0^t \sigma^\alpha(s)[S(0) - \frac{a(s)}{b(s)} + \hat{L}(\hat{T}_s) + \exp[\int_0^s b(u)du]\frac{a(s)}{b(s)}]^\alpha ds$.

8. The Heath, Jarrow, and Morton process driven by α-stable Lévy motion: $f(t,u) = f(0,u) + \int_0^t a(v,u)dv + \int_0^t b(f(v,u))dL(v)$. Solution $f(t,u) = f(0,u) + \hat{L}(\hat{T}_t) + \int_0^t a(v,u)dv$, where $\hat{T}_t = \int_0^t b^\alpha(f(0,u) + \hat{L}(\hat{T}_s) + \int_0^s a(v,u)dv)ds$.

Proof. The approach is to eliminate drift, reduce obtained SDE to the above-mentioned form $dX(t) = a(t,X(t-))dL(t)$, and then use the above-mentioned Rosinski-Woyczynski (1986) result.

8.4.2 Solution of Multifactor Lévy Models Using CTM

Solution of multifactor models driven by α-stable Lévy motions (see Section 8.3.2) can be obtained using various combinations of solutions of the above-mentioned processes and the CTM. We give one example of two-factor continuous-time GARCH model driven by α-stable Lévy motions.

Proposition 2. Let us have the following two-factor Lévy-based model:

$$dS(t) = r(t-)S(t-))dt + \sigma_1 S(t-)dL^1(t)$$
$$dr(t) = a(m - r(t-))dt + \sigma_2 r(t-)dL^2(t),$$

where $L^1 and L^2$ may be correlated, $m \in R, \sigma_i, a > 0, i = 1, 2$.

Then the solution of the two-factor Lévy model using the CTM is (applying CTM for the first and the second equations, respectively)

$$S(t) = e^{\int_0^t r_s ds}[S_0 + \hat{L}^1(\hat{T}_t^1)]$$
$$= e^{\int_0^t e^{-as}[r_0 - m + \hat{L}^2(\hat{T}_s^2)]ds}[S_0 + \hat{L}^1(\hat{T}_t^1)],$$

where $\hat{T}^i$ are defined in 1 and 4, respectively, Section 8.4.1.

Proof. The approach is to eliminate drifts in both equations, reduce the obtained SDEs to the above-mentioned form $dX(t) = a(t,X(t-))dL(t)$, and then to use the above-mentioned Rosinski-Woyczynski (1983) result.

Important Remark. Kallsen and Shiryaev (2002) showed that the Rosinski and Woyczinski (1986) statement cannot be extended to any other Lévy process but α-stable processes. If one considers only non-negative integrands a in $dX(t) = a(t,X(t-))dL(t)$, then we can extend their statement to asymmetric α-stable Lévy processes.

8.5 Applications in Financial and Energy Markets

In this section, we consider various applications of the change of time method for Lévy-based SDEs arising in financial and energy markets: swap and option pricing, interest derivative pricing, and forward and futures contract pricing.

8.5.1 Variance Swaps for Lévy-Based Heston Model

Assume that in the risk-neutral world the underlying asset S_t and the variance follow the following model:

$$\begin{cases} \frac{dS_t}{S_t} = r_t dt + \sigma_t dw_t \\ d\sigma_t^2 = k(\theta^2 - \sigma_t^2)dt + \gamma\sigma_t dL_t, \end{cases}$$

where r_t is the deterministic interest rate, σ_0 and θ are the short and long volatilities, $k > 0$ is a reversion speed, $\gamma > 0$ is a volatility (of volatility) parameter, and w_t and L_t are independent standard Wiener and α-stable Lévy processes ($\alpha \in (0,2]$).
The solution for the second equation has the following form:

$$\sigma^2(t) = e^{-kt}[\sigma_0^2 - \theta^2 + \hat{L}(\hat{T}_t)] + \theta^2,$$

where $\hat{T}_t = \gamma^\alpha \int_0^t [e^{k\hat{T}_s}(\sigma_0^2 - \theta^2 + \hat{L}(\hat{T}_s)) + \theta^2 e^{2k\hat{T}_s}]^{\alpha/2} ds$.

A *variance swap* is a forward contract on annualized variance, the square of the realized volatility. Its payoff at expiration is equal to

$$N(\sigma_R^2(S) - K_{var}),$$

where $\sigma_R^2(S)$ is the realized stock variance (quoted in annual terms) over the life of the contract,

$$\sigma_R^2(S) := \frac{1}{T} \int_0^T \sigma^2(s)ds,$$

K_{var} is the delivery price for variance, and N is the notional amount.

Valuing a variance forward contract or swap is no different from valuing any other derivative security. The value of a forward contract P on future realized variance with strike price K_{var} is the expected present value of the future payoff in the risk-neutral world:

$$P_{var} = E\{e^{-rT}(\sigma_R^2(S) - K_{var})\},$$

where r is the risk-free discount rate corresponding to the expiration date T, and E denotes the expectation.

The realized variance in our case is

$$\sigma_R^2(S) := \frac{1}{T} \int_0^T \sigma^2(s)ds = \frac{1}{T} \int_0^T \{e^{-ks}[\sigma_0^2 - \theta^2 + \hat{L}(\hat{T}_s)] + \theta^2\}ds.$$

The value of the variance swap then is

$$P_{var} = E\{e^{-rT}(\sigma_R^2(S) - K_{var})\}$$
$$= E\{e^{-rT}(\frac{1}{T}\int_0^T\{e^{-ks}[\sigma_0^2 - \theta^2 + \hat{L}(\hat{T}_s)] + \theta^2\}ds - K_{var})\}.$$

Thus, for calculating variance swaps, we need to know only $E\{\sigma_R^2(S)\}$, namely, the mean value of the underlying variance, or $E\{\hat{L}(\hat{T}_s)\}$.

Only moments of order less than α exist for the non-Gaussian family of α-stable distributions. We suppose that $1 < \alpha < 2$ to find $E\{\hat{L}(\hat{T}_s)\}$.

The value of a variance swap for the Lévy-based Heston model is

$$P_{var} = e^{-rT}[\frac{1 - e^{-kT}}{kT}(\sigma_0^2 - \theta^2) + \theta^2 + \frac{\delta T}{2} - K_{var}],$$

where δ is a location parameter.

If $\delta = 0$, then the value of a variance swap for the Lévy-based Heston model is

$$P_{var} = e^{-rT}[\frac{1 - e^{-kT}}{kT}(\sigma_0^2 - \theta^2) + \theta^2 - K_{var}],$$

which coincides with the well-known result by Brockhaus and Long (2000) and Swishchuk (2004).

8.5.2 Volatility Swaps for Lévy-Based Heston Model?

A stock *volatility swap* is a forward contract on the annualized volatility. Its payoff at expiration is equal to

$$N(\sigma_R(S) - K_{vol}),$$

where $\sigma_R(S)$ is the realized stock volatility (quoted in annual terms) over the life of contract,

$$\sigma_R(S) := \sqrt{\frac{1}{T}\int_0^T \sigma_s^2 ds},$$

σ_t is a stochastic stock volatility, K_{vol} is the annualized volatility delivery price, and N is the notional amount

To calculate volatility swaps we need more. From Brockhaus and Long (2000) approximation (which used the second-order Taylor expansion for the function $\sqrt{x}$), we have

$$E\{\sqrt{\sigma_R^2(S)}\} \approx \sqrt{E\{V\}} - \frac{Var\{V\}}{8E\{V\}^{3/2}},$$

where $V := \sigma_R^2(S)$ and $\frac{Var\{V\}}{8E\{V\}^{3/2}}$ is the convexity adjustment.

Thus, to calculate the value of volatility swaps,

$$P_{vol} = \{e^{-rT}(E\{\sigma_R(S)\} - K_{vol})\}$$

we need both $E\{V\}$ and $Var\{V\}$.

For $S\alpha S$ processes only the moments of order $p < \alpha$ exist, $\alpha \in (0, 2]$.

Since the $S\alpha S$ r.v. has "infinite variance", the covariation of two jointly $S\alpha S$ real r.v. with dispersions γ_x and γ_y defined by

$$[X, Y]_\alpha = \frac{E[X|Y|^{p-2}Y]}{E[|Y|^p]} \gamma_y,$$

where $\gamma_y = [Y, Y]_\alpha$ is the dispersion of r.v. Y, has often been used instead of the covariance (and correlation).

One of the possible ways to get volatility swaps for the Lévy-based Heston model is to use covariation.

8.5.3 Gaussian- and Lévy-Based SABR/LIBOR Market Models

SABR model (see Hagan et al. 2002) and the LIBOR Market Model (LMM) (Brace et al. 1997; Piterbarg 2003) have become industry standards for pricing plain-vanilla and complex interest rate products, respectively.

The Gaussian-based SABR model (Hagan et al. 2002) is a stochastic volatility model in which the forward value satisfies the following SDE:

$$\begin{cases} dF_t = \sigma_t F_t^\beta dW_t^1 \\ d\sigma_t = v\sigma_t dW_t^2. \end{cases}$$

In a similar way, we introduce the Lévy-based SABR model, a stochastic volatility model, in which the forward value satisfies the following SDE:

$$\begin{cases} dF_t = \sigma_t F_t^\beta dW_t \\ d\sigma_t = v\sigma_t dL_t, \end{cases}$$

where $L(t)$ is an α-stable Lévy process.

The solution of Lévy-based SABR model using a change of time method has the following expression:

$$F_t = F_0 + \hat{W}(\hat{T}_t^1),$$

$$T_t^1 = \int_0^t \sigma_{T_s^1}^{-2}(F_0 + \hat{W}(s))^{-2\beta} ds,$$

$$\sigma_t = \sigma_0 + \hat{L}(\hat{T}_t^2),$$

$$T_t^2 = v^{-\alpha} \int_0^t (\sigma_0 + \hat{L}(s))^{-\alpha} ds.$$

The expressions for F_t and σ_t give the possibility to calculate many financial derivatives.

8.5.4 Energy Forwards and Futures

Random variables following α-stable distribution with small characteristic exponent are *highly impulsive*, and it is this heavy-tail characteristic that makes this density appropriate for modelling noise which is impulsive in nature, for example, energy prices such as electricity. Here, we introduce two Lévy-based models in energy market: two-factor Lévy-based Schwartz-Smith and three-factor Schwartz. We show how to solve them using the change of time method (see Swishchuk (2009)).

8.5.4.1 Lévy-Based Schwartz-Smith Model

We introduce the Lévy-based Schwartz-Smith model:

$$\begin{cases} \ln(S_t) = \kappa_t + \xi_t \\ d\kappa_t = (-k\kappa_t - \lambda_\kappa)dt + \sigma_\kappa dL_\kappa \\ d\xi_t = (\mu_\xi - \lambda_\xi)dt + \sigma_\xi dW_\xi, \end{cases}$$

where S_t is the current spot price, κ_t is the short-term deviation in prices, and ξ_t is the equilibrium price level.

Let $F_{t,T}$ denote the market price for a futures contract with maturity T and then

$$\ln(F_{t,T}) = e^{-k(T-t)}\kappa_t + \xi_t + A(T-t),$$

where $A(T-t)$ is a deterministic function with explicit expression. We note that κ_t, using change of time for α-stable processes, can be presented in the following form:

$$\kappa_t = e^{-kt}[\kappa_0 + \frac{\lambda_\kappa}{k} + \hat{L}_\kappa(\hat{T}_t)],$$

$$\hat{T}_t = \sigma_\kappa^\alpha \int_0^t (e^{-ks}[\kappa_0 + \frac{\lambda_\kappa}{k} + \hat{L}_\kappa(\hat{T}_s)] - \frac{\lambda_\kappa}{k})^\alpha ds.$$

In this way, the market price for a futures contract with maturity T has the following form:

$$\begin{aligned} \ln(F_{t,T}) &= e^{-kT}[\kappa_0 + \tfrac{\lambda_\kappa}{k} + \hat{L}_\kappa(\hat{T}_t)] \\ &+ \xi_0 + (\mu_\xi - \lambda_\xi)t + \sigma_\xi W_\xi + A(T-t), \end{aligned}$$

where the Lévy process $\hat{L}_\kappa$ and Wiener process W_ξ may be correlated.

If $\alpha \in (1,2]$, then we can calculate the value of Lévy-based futures contracts.

8.5.4.2 Lévy-Based Schwartz Model

We also introduce a Lévy-Based Schwartz model:

$$\begin{cases} d\ln(S_t) = (r_t - \delta_t)S_t dt + S_t\sigma_1 dW_1 \\ \quad d\delta_t = k(a - \delta_t)dt + \sigma_2 dL \\ \quad dr_t = a(m - r_t)dt + \sigma_3 dW_2, \end{cases}$$

where the Wiener processes $W_1 and W_2$ and α-stable Lévy process L may be correlated. δ_t and r_t are the instantaneous convenience yield and interest rate, respectively.

We note that

$$\delta_t = e^{kt}(\delta_0 - a + \hat{L}(\hat{T}_t)),$$
$$\hat{T}_t = \sigma_2^\alpha \int_0^t (e^{ks}[\delta_0 - a + \hat{L}(\hat{T}_s)] + a)^\alpha ds$$

and

$$r_t = e^{at}(r_0 - m + \hat{W}_2(\hat{T}_t)),$$
$$\hat{T}_t = \sigma_3^2 \int_0^t (e^{as}[r_0 - m + \hat{W}_2(\hat{T}_s)] + m)^2 ds.$$

The solution for $\ln[S_t]$

$$\ln[S_t] = e^{\int_0^t [e^{as}(r_0 - m + \hat{W}_2(\hat{T}_s^2)) - e^{ks}(\delta_0 - a + \hat{L}(\hat{T}_s))]ds}[\ln S_0 + \hat{W}_1(\hat{T}_t^1)].$$

In this way, the futures contract has the following form:

$$\begin{aligned} \ln(F_{t,T}) &= \tfrac{1-e^{-k(T-t)}}{k}\delta_t + \tfrac{1-e^{-a(T-t)}}{a}r_t + \ln(S_t) + C(T-t) \\ &= \tfrac{1-e^{-k(T-t)}}{k}[e^{kt}(\delta_0 - a + \hat{L}(\hat{T}_t))] \\ &+ \tfrac{1-e^{-a(T-t)}}{a}e^{at}(r_0 - m + \hat{W}_2(\hat{T}_t^2)) \\ &+ \exp\{\int_0^t (e^{as}(r_0 - m + \hat{W}_2(\hat{T}_s^2)) \\ &- e^{ks}(\delta_0 - a + \hat{L}(\hat{T}_s)))ds\}[\ln(S_0) + \hat{W}_1(\hat{T}_t^1)] \\ &+ C(T-t), \end{aligned}$$

where $C(T-t)$ is a deterministic explicit function. If $\alpha > 1$, then we can calculate the value of a futures contract.

References

Andersen, L. and Andersen, J. Volatility skews and extensions of the Libor Market Model. *Applied Mathematical Finance*, 7:1–32, March 2000.

Andersen, L. and Andersen, J. Volatile volatilities. *Risk,* 15(12), December 2002.

Applebaum, D. *Levy Processes and Stochastic Calculus*, Cambridge University Press, 2003.

Barndorff-Nielsen, E. and Shiryaev, A.N. *Change of Time and Change of Measures*, World Scientific, 2010, 305 p.

Bates, D. Jumps and stochastic volatility: the exchange rate processes implicit in Deutschemark options, *Review Finance Studies*, 9, pp. 69–107, 1996.

Benth, F., Benth, J. and Koekebakker, S. *Stochastic Modelling of Electricity and Related Markets*, World Sci., 2008.

Björk, T. and Landen, C. On the term structure of futures and forward prices. In: Geman, H.,Madan, D., Pliska, S. and Vorst, T., Editors. Mathematical Finance-Bachelier Congress 2000, Springer, Berlin (2002), pp. 111–149.

Black, F. The pricing of commodity contracts, *J. Financial Economics*, 3, 167–179, 1976.

Brace, A., Gatarek, D. and Musiela, M. The market model of interest rate dynamics, *Math. Finance*, 1997, **4**, 127–155.

Brigo, D. and Mercurio, F. *Interest-Rate Models-Theory and Practice.* Springer Verlag, 2001.

Brockhaus, O. and Long, D. Volatility swaps made simple, *RISK*, January, 92–96, 2000.

Carr, P., Geman, H., Madan, D. and Yor, M. Stochastic volatility for Lévy processes.// Mathematical Finance, vol. 13, No. 3 (July 2003), 345–382.

Cartea, A. and Howison, S. Option pricing with Lévy-stable processes generated by Lévy-stable integrated variance. *Birkbeck Working Papers in Economics & Finance*, Birkbeck, University of London, February 24 2006.

Cox, J., Ingersoll, J. and Ross, S. A theory of the term structure of interest rate. *Econometrics,* 53 (1985), pp. 385–407.

Dupire, B. Pricing with a smile, Risk (1999).

Eberlein, E. and Raible, S. Term structure models driven by general Lévy processes, Math. Finance 9(1) (1999), 31–53.

Eydeland, E. and Geman, H. *Pricing power derivatives*, RISK, September, 1998.

Geman, H., Madan, D. and Yor, M. Time changes for Lévy processes, Math. Finance, 11, 79–96, 2001.

Geman, H. and Roncoroni, R. *Understanding the fine structure of electricity prices*, Journal of Business, 2005.

Geman, H. *Scarcity and price volatility in oil markets* (EDF Trading Technical Report), 2000.

Geman, H. *Commodity and Commodity Derivatives: Modelling and Pricing for Agriculturals, Metals and Energy.* Wiley/Finance, 2005.

Gibson and Schwartz, E. *Stochastic convenience yield and the pricing of oil contingent claims*, Journal of Finance, 45, 959–976, 1990.

Girsanov, I. On transforming a certain class of stochastic processes by absolutely continuous substitution of measures. Theory Probab. Appl., 5(1960), 3, pp. 285–301.

Glasserman, P. and Kou, S. The term structure of simple forward rates with jumps risk. Columbia working paper, 1999.

Glasserman, P. and N. Merener. Numerical solution of jump diffusion LIBOR market models, *Fin. Stochastics*, 2003, **7**, 1–27.

Hagan, P., Kumar, D., Lesniewski, A., and Woodward, D. Managing smile risk. *Wilmott Magazine,* autumn, 2002, p. 84–108.

Heath, D., Jarrow, R. and Morton, A. Bond pricing and the term structure of the interest rates: A new methodology. *Econometrica*, 60, 1 (1992), pp. 77–105.

Henry-Laborderè, P. Combining the SABR and LMM models. *Risk,* October, 2007, p. 102–107.

Heston, S. A closed-form solution for options with stochastic volatility with applications to bond and currency options, *Review of Financial Studies*, 6, 327–343, 1993.

Ho, T. and Lee, S. Term structure movements and pricing interest rate contingent claim. *J. of Finance,* 41 (December 1986), pp. 1011–1029.

Hull, J. and White, A. Pricing interest rate derivative securities. *Review of Fin. Studies*, 3,4 (1990), pp. 573–592.

Jacod, J. *Calcul Stochastique et Problèmes de Martingales*, Lecture Notes in Mathematics 714, Springer-Verlag, 1979.

Jamshidian, F. LIBOR and swap market models and measures, *Fin. Stochastics*, 1 (4), 293–330, 1997.

Janicki, A., Michna, Z., and Weron, A. Approximation for SDEs driven by α-stable Lévy motion, Appl. Mathematicae 24 (1996), 149–168.

Joshi, M. and Rebonato, R. A displaced-diffusion stochastic volatility Libor market model: motivation, definition and implementation. *Quantitative Finance*, 3, 2003, p. 458–469.

Kallenberg, O. Some time change representations of stable integrals, via predictable transformations of local martingales. *Stochastic Processes and Their Applications,* 40 (1992), 199–223.

Kallsen, J. and Shiryaev, A. Time change representation of stochastic integrals, *Theory Probab. Appl.,* vol. 46, N. 3, 522–528, 2002.

Lévy, P. *Processus Stochastiques et Mouvement Brownian,* 2nd ed., Gauthier-Villars, Paris, 1965.

Madan, D. and Seneta, E. The variance gamma (VG) model for share market returns, J. Business 63, 511–524, 1990.

Ornstein, L. and G.Uhlenbeck. On the theory of Brownian motion. *Physical Review,* 36 (1930), 823–841.

Pilipovic, D. *Valuing and Managing Energy Derivatives*, New York, McGraw-Hill, 1997.

Piterbarg, V. Astochastic volatility forward Libor model with a term structure of volatility smiles. October, 2003 (http://ssrn.com/abstract=472061)

Piterbarg, V. Time to smile. *Risk*, May 2005, p. 87–92.

Protter, P. *Stochastic Integration and Differential Equations*, Springer, 2005.

Rebonato, R. A time-homogeneous, SABR-consistent extension of the LMM, *Risk*, 2007.

Rebonato, R. *Modern pricing of interest rate derivatives: the Libor market model and beyond*. Princeton University Press, 2002.

Rebonato, R. and Joshi, M. A joint empirical and theoretical investigation of the models of deformation of swaption matrices: implications for the stochastic-volatility Libor market model. Intern. J. Theoret. Applied Finance, 5(7), 2002, p. 667–694.

Rebonato, R. and Kainth, D. A two-regime, stochastic-volatility extension of the Libot market model. Intern. J. Theoret. Applied Finance, 7(5), 2004, p. 555–575.

Rosinski, J. and Woyczinski, W. On Ito stochastic integration with respect to *p*-stable motion: Inner clock, integrability of sample paths, double and multiple integrals, *Ann. Probab.*, 14 (1986), 271–286

Sato, K. *Lévy Processes and Infinitely Divisible Distributions.* Cambridge University Press, Cambridge, UK, 1999.

Schoutens, W. *Lévy Processes in Finance. Pricing Financial Derivatives.* Wiley & Sons, 2003.

Schwartz, E. The stochastic behaviour of commodity prices: implications for pricing and hedging, *J. Finance*, 52, 923–97, 1997.

Sin, C. Alternative interest rate volatility smile models. *Risk conference proceedings*, 2002.

Schwartz, E. Short-Term Variations and Long-Term Dynamics in Commodity Prices, *Management Science*, Volume 46, Issue 7, 1990.

Schwartz, E. The stochastic behaviour of commodity prices: implications for pricing and hedging, *J. Finance*, **52**, 923–973, 1997.

Shiryaev, A. *Essentials of Stochastic Finance*, World Scientific, 2008.

Skorokhod, A. *Random Processes with Independent Increments*, Nauka, Moscow, 1964. (English translation: Kluwer AP, 1991).

Swishchuk, A. Lévy-based interest rate derivatives: change of time and PIDEs, submitted to CAMQ, June 10, 2008 (available at http://papers.ssrn.com/sol3/papers.cfm?abstract_id=1322532).

Swishchuk, A. Multi-factor Lévy models for pricing of financial and energy derivatives, *CAMQ*, V. 17, No. 4, Winter, 2009

Swishchuk, A. Modelling and valuing of variance and volatility swaps for financial markets with stochastic volatilities, *Wilmott Magazine*, Technical Article, N0. 2, September, 2004, 64–72.

Swishchuk, A. Change of time method in mathematical finance, *Canad. Appl. Math. Quart.*, vol. 15, No. 3, 2007, p. 299–336.

Swishchuk, A. Explicit option pricing formula for a mean-reverting asset in energy market, J. of Numer. Appl. Math., Vol. 1(96), 2008, pp. 216–233.

Vasicek, O. An equilibrium characterization of the term structure. *J. of Finan. Economics*, 5 (1977), pp. 177–188.

Villaplana A *two-state variables model for electricity prices*, Third World Congress of the Bachelier Finance Society, Chicago, 2004.

Wilmott, P. *Paul Wilmott on Quantitative Finance*, New York, Wiley, 2000.

Zanzotto, A. On solutions of one-dimensional SDEs driven by stable Lévy motion, Stoch. Process. Appl. 68 (1997), 209–228.

Epilogue

"Time you enjoy wasting, was not wasted". —John Lennon.

The present book was devoted to the history of change of time methods (CTM), connection of CTM with stochastic volatilities and finance, and many applications of CTM. As a reader may noticed, this book is a brief introduction to the theory of CTM and may be considered as a handbook in this area.

I hope that you enjoyed your time while reading this book and thank you for getting to this very end.

If you have any comments, remarks, etc., please send them to:

Anatoliy Swishchuk
University of Calgary
Department of Mathematics and Statistics
2500 University Drive NW
Calgary, AB,Canada T2N 1N4
E-mail: aswish@ucalgary.ca

© The Author 2016   125
A. Swishchuk, *Change of Time Methods in Quantitative Finance*,
SpringerBriefs in Mathematics, DOI 10.1007/978-3-319-32408-1

Index

© The Author 2016
A. Swishchuk, *Change of Time Methods in Quantitative Finance*,
SpringerBriefs in Mathematics, DOI 10.1007/978-3-319-32408-1

Printed in the United States
By Bookmasters